Guideline No. 31

Fresh prepared produce: GMP for high oxygen MAP and non-sulphite dipping

Compiled by Brian P.F. Day

2001

Campden & Chorleywood Food Research Association Group comprises
Campden & Chorleywood Food Research Association
and its subsidiary companies
CCFRA Technology Ltd   CCFRA Group Services Ltd   Campden & Chorleywood Magyarország, Hungary

Campden & Chorleywood Food
Research Association Group

Chipping Campden,
Gloucestershire
GL55 6LD UK
Tel: +44 (0) 1386 842000
Fax: +44 (0) 1386 842100
www.campden.co.uk

Information emanating from this company is given after the exercise of all reasonable care and skill in its compilation, preparation and issue, but is provided without liability in its application and use.

Legislation changes frequently. It is essential to confirm that legislation cited in this publication and current at the time of printing, is still in force before acting upon it.

The information contained in this publication must not be reproduced without permission from the Publications Manager.

© CCFRA 2001
ISBN: 0 905942 36 1

EXECUTIVE SUMMARY

This guidelines document provides users of novel high oxygen modified atmosphere packaging (MAP) technology with clear and concise practical advice and good manufacturing practice (GMP) recommendations on how to successfully apply this technology to fresh prepared fruit and vegetables (i.e. produce), in conjunction with appropriate non-sulphite dipping treatments. It provides important background information, explains the rationale behind novel MAP and non-sulphite dipping and summarises all the important results from a major European Union (EU) and industry funded project on "Novel MAP for fresh prepared produce". The document contains concise practical advice and recommendations (e.g. safety of high oxygen MAP, optimal high oxygen mixtures, produce volume/gas ratios, packaging materials, chilled storage temperatures, non-sulphite formulations and dipping protocols) in order to facilitate commercial exploitation. Brief reference in the document has been made with respect to novel argon and nitrous oxide MAP, but in light of the variable results obtained for these novel MAP treatments, the majority of the text concentrates on the applications of novel high oxygen MAP.

ACKNOWLEDGEMENTS

CCFRA gratefully acknowledges the financial support of the EU FAIR programme and the contributions of CCFRA's EU FAIR project partners for the underpinning research findings described in section 4 of this document. The EU FAIR CT96-1104 project on "Novel MAP for fresh prepared produce" (September, 1996 - August, 1999) was coordinated by CCFRA and also involved the following partner organisations:

Agrotechnological Research Institute (ATO-DLO), The Netherlands
Swedish Institute for Food and Biotechnology (SIK), Sweden
Technical Research Centre of Finland (VTT), Finland
University of Limerick (UL), Ireland
Istituto Nazionale della Nutrizione (INN), Italy

CCFRA also gratefully acknowledges the financial and technical support of members of two industrially funded research Clubs that were set up to help investigate the interesting effects of novel MAP on fresh prepared produce. CCFRA's High Oxygen MAP Club (April, 1995 - September, 1997) and Novel Gases MAP Club (January, 1998 - December, 1999) comprised the following companies:

AGA AB, Sweden	(Apr 1995 - Dec 1999)
Air Liquide SA, France	(Jun 1997 - Dec 1999)
Air Products PLC, UK	(Apr 1995 - Dec 1999)
BOC Gases Europe, UK	(Apr 1995 - Sep 1997)
Danisco Flexibles P-Plus, UK	(Jan 1998 - Dec 1998)
EPL Technologies, Inc., USA	(Apr 1995 - Dec 1999)
Fisher Chilled Foods plc, UK	(Sep 1995 - Sep 1997)
Geest Prepared Produce, UK	(Apr 1995 - Jun 1999)

Global Fresh Technology, Inc., USA   (Jan 1996 - Jun 1996)
Grace Italiana spa., Italy   (Jan 1996 - Dec 1996)
Hitchen Foods plc, UK   (Apr 1995 - Sep 1997)
Hitech Instruments Ltd., UK   (Jan 1998 - Dec 1999)
Kanes Foods Ltd., UK   (Apr 1995 - Mar 1996)
Kettle Produce Ltd., UK   (Apr 1995 - Dec 1999)
Lawson Mardon Flexibles, UK   (Apr 1995 - Mar 1996)
MAP Systems UK Ltd., UK   (Jan 1998 - Dec 1999)
Marks & Spencer plc, UK   (Apr 1995 - Dec 1998)
Messer UK Ltd., UK   (Apr 1995 - Sep 1997)
Multivac UK Ltd., UK   (Apr 1995 - Mar 1996)
Rexam H.P. Flexibles, UK   (Jan 1996 - Mar 1999)
Safeway Stores plc, UK   (Apr 1995 - Dec 1996)
Sainsburys Supermarkets Ltd., UK   (Apr 1995 - Sep 1997)
Southern Salads Ltd., UK   (Jan 1998 - Dec 1999)
Tinsley Foods plc, UK   (Jan 1998 - Dec 1999)
Witt Gas Techniques Ltd., UK   (Jan 1998 - Dec 1999)

SCOPE AND DOCUMENT FORMAT

The objectives of this guidelines document are to provide the prepared produce and related industries with advice and recommendations to assure the safety and extended shelf-life of a wide range of high oxygen (O_2) modified atmosphere (MA) packed fresh prepared produce items. In addition, guidance is provided on suitable protocols for non-sulphite dipping treatments for certain fresh prepared produce items.

The contents of this document reflect a consensus of informed technical opinion from the prepared produce, retail, food service, packaging film, packaging machinery, instrument control, gas supply and non-sulphite dipping supply industries. In addition, underpinning background research results from CCFRA's EU FAIR partners are summarised and appropriate advice inserted into specific sections of this guidelines document.

It should be appreciated that the applications of high O_2 MAP technology and non-sulphite dipping treatments are recent innovations and new knowledge will evolve rapidly in the future. Hence, the contents of this guidelines document only reflect the present status of development of high O_2 MAP and non-sulphite dipping treatments and currently available knowledge. Notwithstanding, current gaps in knowledge and possible research directions are included, in order to assist researchers in the future.

The advice and recommendations provided by this document are only voluntary, but are intended to supplement, assist and encourage compliance with applicable EU Directives and national food laws and regulations in European Member States. Emphasis is placed specifically on high O_2 MAP and non-sulphite dipping of fresh prepared produce items. To avoid unnecessary overlap, extensive references are made to published documents which cover more general advice and recommendations for good manufacturing practices of chilled perishable foods and related relevant subject areas.

The advice and recommendations in this document only cover applications to fresh prepared produce items. Applications of high O_2 MAP to chilled combination food items (e.g. ready meals, pizzas, kebabs and sandwiches) are outside the scope of this document but are currently the subject of on-going research (Day et al, 1999; Day, 2000b).

The contents of this document are arranged to include relevant background information, the rationale behind novel MAP and non-sulphite dipping, guidelines for high O_2 MAP and non-sulphite dipping, a summary of major research findings and future directions, references and several appendices. Extensive references to other currently available documents are made to complement the specific guidance outlined in this document.

CONTENTS

			Page No.
1.	BACKGROUND INFORMATION		1
	1.1	Health benefits of fruit and vegetable consumption	1
	1.2	Fresh prepared produce	2
	1.3	Spoilage and shelf-life of fresh prepared produce	3
	1.4	MAP of fresh prepared produce	4
2.	RATIONALE BEHIND NOVEL MAP AND NON-SULPHITE DIPPING		7
	2.1	High O_2 MAP	7
	2.2	Argon and nitrous oxide MAP	9
	2.3	Non-sulphite dipping	10
3.	GUIDELINES FOR HIGH O_2 MAP AND NON-SULPHITE DIPPING		12
	3.1	Introduction and general guidance	12
		3.1.1 Produce raw materials	13
		3.1.2 Hygiene and handling	14
		3.1.3 Temperature control	15
		3.1.4 Preparation procedures	16
		3.1.4.1 Trimming and peeling	16
		3.1.4.2 Cutting and slicing	18
		3.1.4.3 Washing and decontamination	18
		3.1.4.4 Dewatering	21
		3.1.5 Quality assurance tests	21
		3.1.6 Labelling	22
	3.2	Non-sulphite dipping	22
		3.2.1 Produce raw materials	23
		3.2.2 Pre-dipping preparation treatments	23
		3.2.3 Dipping procedures	24
		3.2.4 Post-dipping treatments	25

			Page No.
	3.3	High O_2 MAP	26
		3.3.1 Safety	26
		3.3.2 Optimal gas levels	27
		3.3.3 Produce volume/gas volume ratio	29
		3.3.4 Packaging materials	30
		3.3.5 Temperature control	32
		3.3.6 Fresh prepared produce applications	33
4.	CCFRA CLUB AND EU FAIR FUNDED RESEARCH		35
	4.1	Objectives	35
	4.2	Research tasks	35
	4.3	Summary of major findings	36
		4.3.1 Safety of high O_2 MAP	36
		4.3.2 Non-sulphite dipping and novel MAP	38
		4.3.3 Effects of novel MAP on fresh prepared produce	39
		4.3.4 Effects of novel MAP on microbial growth	42
		4.3.5 Effects of novel MAP on biochemical reactions	45
		4.3.6 Effects of high O_2 on nutritional components	46
5.	FUTURE RESEARCH DIRECTIONS		48
6.	REFERENCES		51
	APPENDICES		
	Appendix I	Reported respiration rates of whole and prepared fresh produce items in air	60
	Appendix II	Oxygen and water vapour transmission rates of selected packaging materials for fresh produce	68
	Appendix III	Graded sensory quality specification for fresh prepared iceberg lettuce	69
	Appendix IV	Summary tables of the results from CCFRA's fresh prepared produce trials	70

1. BACKGROUND INFORMATION

1.1 Health benefits of fruit and vegetable consumption

The consumption of fresh produce (i.e. fruit and vegetables) has long been associated with a healthy diet (Serafini, 1999). Overwhelming epidemiological evidence exists to demonstrate that increased consumption of fresh produce is correlated with a lower risk of most cancers, coronary heart disease (CHD) and other degenerative diseases (e.g. Block *et al*, 1992; Ness and Powles, 1997). For example, the greater consumption of fresh produce by Southern Europeans in the Mediterranean areas is considered to be largely responsible for their lower mortality rates from cancer, CHD and atherosclerosis compared with Northern Europeans (Ferro-Luzzi *et al*, 1999; Serafini, 1999). The importance of a healthy diet is recognised by national governments and international organisations who have dedicated much effort through nutritional policies and educational campaigns to encourage the increased consumption of fresh produce.

Fresh produce contains an array of bioactive compounds which are thought to play an important role in the prevention of several chronic diseases (Puupponen-Pimiä, 1999). Their activity appears to be linked to their capacity to modulate in-vivo oxidative stress caused by an imbalance between reactive oxygen species (ROS) and antioxidant defences. ROS cause oxidative damage to biological macromolecules such as DNA, lipids, carbohydrates and proteins, which can lead to the initiation and progression of several chronic diseases. Diets rich in fresh produce which contain powerful antioxidants (e.g. vitamins C and E, carotenoids, phenolics and flavonoids) play an essential role in health protection by minimising in-vivo oxidative stress and consequent oxidative damage to biological macromolecules (Ferro-Luzzi *et al*, 1999; Serafini, 1999). Fresh produce also contain many other compounds (e.g. minerals, omega-3 fatty acids, dietary fibre, prebiotic oligosaccharides, phytoestrogens, phytosterols, glucosinolates and bioactive peptides and proteins) with either established or potential health benefits. All the various components of fresh produce are currently the target of intensive international research which has already led to the development of beneficial functional foods and a better understanding of their nutritional properties and importance in a healthy diet (Puupponen-Pimiä, 1999).

1.2 Fresh prepared produce

Many synonyms are used for the term "fresh prepared" produce, and these include "lightly processed", "minimally processed", "partially processed", "fresh processed", "pre-cut" and "fresh-cut" produce. All of these synonyms refer to fresh produce that has been subjected to processing operations such as cutting, slicing, shredding, peeling, trimming, coring or chopping, that do not affect the "fresh-like" characteristics of the produce. Fresh prepared produce is not subjected to any heat or freezing treatments but usually receives washing/decontamination, dipping and/or dewatering treatments before packaging and chilled storage, distribution and marketing (Schlimme, 1995).

During recent years there has been an explosive growth in the market for fresh prepared produce items in Western Europe and North America. In the UK, the fresh produce market was estimated to be worth £6.42 billion in 1999 and has been forecasted to be worth £7.32 billion in 2004 (Keynote, 2000). Fresh prepared produce has been accounting for an increasing proportion of the total fresh produce market and this proportion has been estimated to be growing at an annual rate of 10-25% per annum since 1990. In the USA, the market share of fresh prepared produce was estimated to account for 8-10% of total produce sales in 1996 and has been forecasted to account for an increasing proportion of all produce sales in the future (Day and Gorris, 1993; Garrett, 1994; Cooper, 1999). Many other countries have only recently introduced fresh prepared produce into the retail marketplace but considerable growth potential in the near future is anticipated (Garrett, 1998). The main driving force for this market growth is the increasing consumer demand for fresh, healthy, convenient and additive-free prepared produce items which are safe and nutritious (CCFRA, 1999b). In addition, the benefits of adding value to whole produce commodities and reducing wastage and costs have been very attractive to the prepared produce, food service and retail industries.

1.3 Spoilage and shelf-life of fresh prepared produce

All fresh prepared produce items are highly perishable and more perishable than unprocessed whole fresh produce as a consequence of the tissue damage resulting from minimal processing operations (Schlimme, 1995). Hence, since fresh prepared produce is essentially wounded, shelf-life is often greatly diminished and can be as short as 1-3 days but typically is 4-10 days at chilled temperatures of storage (Burns, 1995; Ahvenainen, 1996; Garrett, 1998). Loss of cellular integrity at the cut surface of fresh prepared produce destroys the compartmentalisation of intracellular enzymes and substrates which are liberated, mixed and cause undesirable physiological changes (Miller, 1992; Brecht, 1995). Deleterious enzymic discoloration reactions and formation of off-odours and flavours are often the consequence as respiration and ethylene production rates increase as a response to tissue wounding. Also, exudate released from the cut surface of fresh prepared produce is a favourable medium for microbial growth. Minimising the negative consequences of tissue wounding in fresh prepared produce can extend shelf-life by helping to maintain the sensory, microbial and nutritional qualities of fresh prepared produce (Brecht, 1995).

The principal spoilage mechanisms affecting the quality and restricting the shelf-life of fresh prepared produce are enzymic discoloration, microbial growth and moisture loss (Day, 1994). Extensive reviews are available which describe these spoilage mechanisms in detail as well as outlining methods to inhibit them and improve the microbial safety of fresh prepared produce (Day, 1994; Nguyen-the and Carlin, 1994; Wiley, 1994; Hurst, 1995; Romig, 1995; Ahvenainen, 1996; Francis *et al*, 1999; Martens, 1999; Seymour, 1999; Heard, 2000). In addition, good manufacturing and handling practices along with proper chilled temperature control and the appropriate use of MAP are also very effective at inhibiting these spoilage mechanisms, thereby extending the shelf-life of fresh prepared produce items (Day, 1992; IFPA, 1993; Betts 1996; IFPA, 1996; CFA, 1997; WHO, 1998; Chambers, 1999; CHGL, 1999; IFPA, 1999; Knight, 1999; Seymour, 1999; IFPA, 2000). Shelf-life extension also results in the commercial benefits of less wastage in manufacturing, longer distribution channels, improved image and the ability to sell convenient, fresh prepared produce items to the consumer with reasonable remaining chilled storage time (Day, 1994).

1.4 MAP of fresh prepared produce

MAP has become a widely used food preservation technique which minimally affects fresh product characteristics, and hence fits in well with the recent consumer preference for fresh, convenient and additive-free foods. MAP is now increasingly being used for extending the shelf-life of a wide range of fresh prepared produce items (Day, 1998).

Unlike other chilled perishable foods, fresh produce continues to respire after harvesting, and consequently any subsequent packaging must take into account this respiratory activity. Respiration is a very complex biological process whereby carbohydrates, polysaccharides, organic acids and other energy sources are metabolised into simpler molecules with the production of heat. The end products of aerobic respiration are carbon dioxide (CO_2) and water, whereas undesirable fermentation products such as ethanol, acetaldehyde and ketones are produced during anaerobic respiration (Kader *et al*, 1989).

The depletion of O_2 and enrichment of CO_2 are natural consequences of the progress of respiration when fresh whole or prepared produce items are stored in hermetically sealed packs. Such modification of the atmosphere results in an aerobic respiratory rate decrease with a consequent extension of shelf-life (Kader *et al*, 1989). However, excessive depletion of O_2 levels to <2% O_2 will initiate a substantial rise in the anaerobic respiratory rate with a consequent adverse effect on achievable shelf-life (Day, 1994).

MAs can passively evolve within hermetically air-sealed packs as a consequence of produce respiration. If a produce item's respiratory rate (Appendix I) is properly matched to packaging film of appropriate gas permeability (Appendix II), then a beneficial equilibrium MA (EMA) can be passively established. Generally speaking, packaging films of high gas permeability are required for highly respiring produce items, whereas lower gas permeability packaging films can be used for low respiring items (Exama *et al*, 1993; Day, 1994; Marston, 1995; Garrett, 1998; O'Beirne, 1999).

However, in the MAP of fresh produce, there is a limited ability to regulate passively established MAs within hermetically air-sealed packs. There are many circumstances when it is desirable to rapidly establish the atmosphere within produce packs. By replacing the pack atmosphere with a desired mixture of O_2, CO_2 and nitrogen (N_2), a beneficial EMA may be established more rapidly than a passively generated EMA. For example, flushing packs with N_2 (to achieve a residual O_2 level of 5-10%) or a mixture of 5-10% O_2, 5-10% CO_2 and 80-90% N_2 is commercial practice for inhibiting undesirable browning and pinking on prepared leafy green salad vegetables (Day, 1998; Garrett, 1998).

At the present time, the key to successful MAP of fresh prepared produce is to use packaging film of appropriate gas permeability so as to establish optimal EMAs of typically 2-10% O_2 and 10-20% CO_2 (Marston, 1995). Exposure of fresh prepared produce to O_2 and CO_2 levels outside these ranges can initiate undesirable anaerobic respiration as well as causing physiological disorders. Different produce items vary considerably in their tolerances to low O_2 and high CO_2 levels and classification of whole produce commodities according to their tolerance to different atmospheres has been extensively reviewed (Kader *et al*, 1989). It should be appreciated that fresh prepared produce items can tolerate lower O_2 levels and higher CO_2 levels than their corresponding whole produce commodity (Day, 1994). Hence, optimal EMAs reduce respiration rates without causing physiological damage to fresh prepared produce. Establishment of an optimal EMA can extend fresh prepared produce shelf-life by delaying ripening, decreasing ethylene production and sensitivity, retarding textural softening, reducing chlorophyll degradation and enzymic discoloration, preserving vitamins and retarding microbial growth (Kader *et al*, 1989).

Establishment of an EMA inside a fresh prepared produce pack is influenced by numerous variables, such as respiration rate (which itself is affected by temperature; atmospheric composition; produce type, variety, cultivar and maturity; and severity of preparation); packaging film gas permeability; pack volume, surface area and fill weight; produce volume/gas volume ratio and degree of illumination (Kader *et al*, 1989; Day, 1994; O'Beirne, 1999). The influence of all these variables needs to be considered along with an understanding of the intrinsic nature of the prepared produce (e.g. pH, water content, biological structure and ethylene production and sensitivity) in order to gain the full benefits from MAP (Day, 1994). In addition, various extrinsic factors (e.g. harvesting, handling, hygiene and temperature control) need to be optimised (IFPA, 1999).

Consequently, establishment of an optimum EMA for individual produce items is very complex. Furthermore, in many commercial situations, produce is sealed in packaging film of insufficient gas permeability (IFPA, 1993; Betts, 1996) resulting in development of undesirable fermentation reactions and potentially hazardous anaerobic conditions (e.g. <2% O_2 and >20% CO_2). Recently developed microperforated films, which have very high gas transmission rates, are now commercially used for maintaining aerobic EMAs (e.g. 5-15% O_2 and 5-15% CO_2) for highly respiring prepared produce items such as broccoli and cauliflower florets, baton carrots, beansprouts, mushrooms and spinach (Geeson, 1999). However, microperforated films are more expensive than conventional plastic films and potentially allow for the ingress of microorganisms into sealed packs during wet handling situations (Rose, 1994). In addition, microperforated films can allow moisture to escape from certain sensitive prepared produce items (e.g. baton carrots) to the outside of the pack. Also, certain odourous produce items (e.g. prepared onions and garlic) can taint other food products if packed in microperforated film (Day, 1998).

Commercially, most light prepared salad items are MA packed on vertical form-fill-seal (VFFS) and horizontal form-fill-seal (HFFS) machines (Hartley, 2000). These machines use a gas flushing technique to introduce gas into MA pillow-packs. It should be noted that thermoform-fill-seal (TFFS) machines are commercially used for MA packaging of heavier and/or delicate prepared produce items and there are no technical reasons why they cannot be used for light prepared salad items. TFFS machines use a compensated vacuum technique to evacuate air and then introduce gas into tray and lidding film MA packs, and have been safely used for many years for the retail MAP of red meats under high O_2 levels (60 - 80% O_2). Other commercially available MAP machines include vacuum chamber (VC), preformed tray and lidding film (PTLF) and snorkel type (ST) machines (Day, 1992).

2. RATIONALE BEHIND NOVEL MAP AND NON-SULPHITE DIPPING

2.1 High O_2 MAP

Information gathered by the author during 1993-1994 revealed that a few fresh prepared produce companies had been experimenting with high O_2 (e.g. 70-100%) MAP and had achieved some surprisingly beneficial results (Day, 1998). High O_2 MAP of prepared produce was not exploited commercially during that period, probably because of the inconsistent results obtained, a lack of understanding of the basic biological mechanisms involved and concerns about possible safety implications. Intrigued by the concept of high O_2 MAP, CCFRA carried out limited experimental trials on prepared iceberg lettuce and tropical fruits, in early 1995. The results of these trials confirmed that high O_2 MAP could overcome the many disadvantages of low O_2 MAP. High O_2 MAP was found to be particularly effective at inhibiting enzymic discolorations, preventing anaerobic fermentation reactions and inhibiting microbial growth. In addition, the high O_2 MAP of prepared produce items within inexpensive, hermetically sealed plastic films was found to be very effective at preventing undesirable moisture and odour losses and ingress of microorganisms during wet handling situations (Day, 1998).

The experimental finding that high O_2 MAP is capable of inhibiting aerobic and anaerobic microbial growth can be explained by the schematic growth profiles of aerobes and anaerobes (Figure 1). By definition, anaerobes grow best under very low O_2 levels and therefore anaerobes would be inhibited under high O_2 conditions. In contrast, aerobes grow best under an atmospheric O_2 level (21%). Hence, under reduced or elevated O_2 levels, there would be growth inhibition of aerobic microorganisms. Under high O_2 MAP, it is hypothesised that reactive oxygen species damage vital cellular macromolecules and thereby inhibit microbial growth when oxidative stresses overwhelm cellular antioxidant protection systems (Gonzalez Roncero and Day, 1998; Serafini, 1999; Kader and Ben-Yehoshua, 2000). Also intuitively, high O_2 MAP inhibits undesirable anaerobic fermentation reactions (Day, 1998).

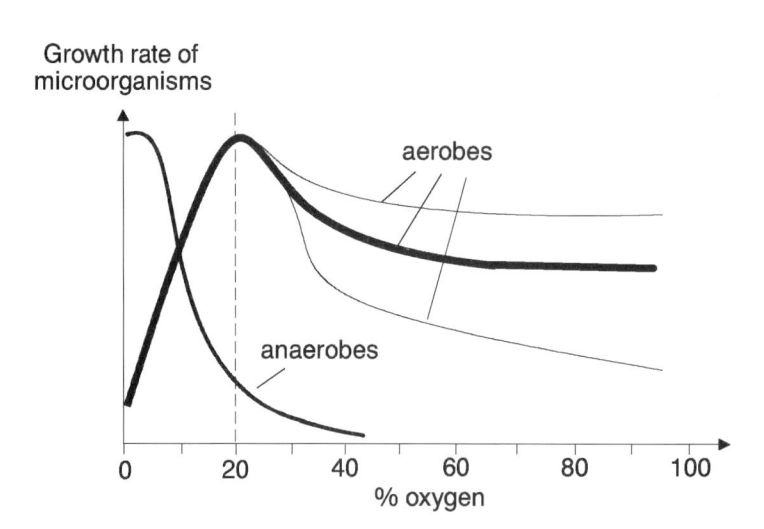

Figure 1: Hypothesised inhibition of microbial growth by high O_2 MAP

Polyphenol oxidase (PPO) is the enzyme primarily responsible for initiating discoloration on the cut surfaces of fresh prepared produce. PPO catalyses the oxidation of natural phenolic substances to colourless quinones which subsequently polymerise to coloured melanin-type compounds (McEvily *et al*, 1992; Duncan, 1999). It is hypothesised that high O_2 levels may cause substrate inhibition of PPO or alternatively, high levels of colourless quinones subsequently formed (Figure 2) may cause feedback product inhibition of PPO (Day, 1998).

Figure 2: Hypothesised inhibition of enzymic discoloration by high O_2 MAP

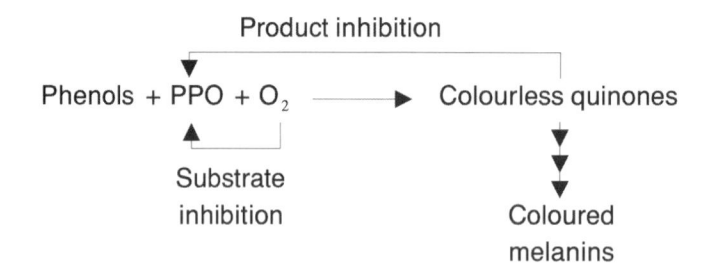

2.2 Argon and nitrous oxide MAP

Argon (Ar) and nitrous oxide (N_2O) are classified as miscellaneous additives and are permitted gases for food use in the European Union (EU). Air Liquide S.A. (Paris, France) has stimulated recent commercial interest in the potential MAP applications of using Ar and, to a lesser extent, N_2O. Air Liquide's broad range of patents claim that in comparison with N_2, Ar can more effectively inhibit enzymic activities, microbial growth and degradative chemical reactions in selected perishable foods (Brody and Thaler, 1996; Spencer, 1999). More specifically, an Air Liquide patent for fresh produce applications claims that Ar and N_2O are capable of extending shelf-life by inhibiting fungal growth, reducing ethylene emissions and slowing down sensory quality deterioration (Fath and Soudain, 1992). Of particular relevance is the claim that Ar can reduce the respiration rates of fresh produce and hence have a direct effect on extension of shelf-life (Spencer, 1999).

Although Ar is chemically inert, Air Liquide's research has indicated that it may have biochemical effects, probably due to its similar atomic size to molecular O_2 and its higher solubility in water and density compared with N_2 and O_2. Hence, Ar is probably more effective at displacing O_2 from cellular sites and enzymic O_2 receptors with the consequence that oxidative deterioration reactions are likely to be inhibited. In addition, Ar and N_2O are thought to sensitise microorganisms to antimicrobial agents. This possible sensitisation is not well understood but may involve alteration of the membrane fluidity of microbial cell walls with a subsequent influence on cell function and performance (Thom and Marquis, 1984). Clearly, more independent research is needed to better understand the potential beneficial effects of Ar and N_2O (Day, 1998).

2.3 Non-sulphite dipping

Enzymic discoloration of fresh prepared produce is one of the major causes of quality loss and spoilage during post-harvest handling, processing and storage (Sapers, 1993; Laurila *et al*, 1998). PPO (EC 1.10.3.1) is the enzyme primarily responsible for the discoloration of fresh prepared potatoes, apples, carrots, parsnips, swede, pears, mushrooms, bananas, peaches, grapes and lettuce, and this discoloration is often the shelflife-limiting quality attribute for these items (Duncan, 1999). PPO activity also results in detrimental changes to the texture and flavour of fresh prepared produce and losses of nutritional quality (Whitaker, 1996).

Given the deleterious effects of PPO activity upon the sensory and nutritional quality of fresh prepared produce, it is not surprising that considerable research has been devoted to inhibit the activity of this enzyme (Duncan, 1999). Sulphites have long been used as food additives to inhibit enzymic and non-enzymic discolorations; to control the growth of microorganisms; and to act as bleaching agents and antioxidants (McEvily *et al*, 1992; Sapers, 1993; Laurila *et al*, 1998). The most frequently used sulphiting agents for fresh prepared produce are sodium and potassium bisulphites and metabisulphites. Sulphites act as PPO inhibitors and antimicrobial agents, and are most effective in acidic conditions (e.g. pH 3-5). For low-acid (e.g. pH 5-8) fresh prepared produce items such as mushrooms, bananas, potatoes and lettuce, sulphites have the tendency to accelerate bacterial decay by adversely affecting cell wall or membrane integrity which may stimulate the growth of certain spoilage bacteria (Duncan, 1999). Also, there are several negative attributes associated with sulphite use which has led to decreased consumer acceptance (McEvily *et al*, 1992). In particular, sulphites can induce severe allergic reactions or even anaphylactic shock in a proportion of the asthmatic population (Sapers, 1993). Consequently, the adverse health effects of sulphite consumption have resulted in stricter regulatory restrictions and consumer labelling requirements (Anon., 1991).

The increased regulatory restrictions on the use of sulphites have created an urgent need for safe, practical and functional alternatives which are economically viable (Ahvenainen, 1996). Proprietary chemical non-sulphite formulations (containing for example, mixtures of ascorbic acid or erythorbic acid or their sodium salts in combination with citric acid, malic acid, tartaric acid, succinic acid, calcium chloride, sodium chloride, 4-hexylresorcinol, sodium acid pyrophosphate and/or cysteine hydrochloride) are commercially available but further research is required to optimise appropriate formulations and dipping protocols for fresh prepared produce items. New opportunities exist for the use of approved starch and pectin-based edible coatings and safe biological agents such as enzymes and PPO inhibitors produced by lactic acid bacteria (McEvily *et al*, 1992; Ahvenainen, 1996; Laurila *et al*, 1998).

It should be appreciated that different produce cultivars show large differences in their tendency to discolour after tissue wounding upon preparation. Such differences can be exploited by selecting raw material cultivars that have a low tendency to discolour after preparation so that treatments to inhibit enzymic discoloration can be minimised (Sapers, 1993). In addition, recent research has demonstrated that combining chemical non-sulphite dipping treatments with optimal MAP yields extended shelf-life and quality benefits greater than those achieved with either dipping or MAP alone (Duncan, 1999). Such combination treatments are likely to be the focus of future research aimed at minimising enzymic discolorations and maximising the maintenance of fresh prepared produce quality.

3. GUIDELINES FOR HIGH O₂ MAP AND NON-SULPHITE DIPPING

3.1 Introduction and general guidance

There are many important factors that impinge on the safety and quality of fresh prepared produce, and all of these factors need to be recognised (see sections 1.3 and 1.4). Table 1 lists some of the key requirements, from "farm to fork", for maintaining the sensory quality and assuring the microbial safety of fresh prepared produce items (Ahvenainen, 1996).

Table 1: Key requirements for fresh prepared produce

- High quality raw materials (correct cultivar variety, correct cultivation, harvesting and storage conditions)
- Strict hygiene, handling and good manufacturing practices; use of hazard analysis and critical control points (HACCP) principles
- Low chilled temperatures during processing
- Careful cleaning and/or washing before and after preparation
- Good quality water (sensory, microbiology, pH) for washing
- Use of safe and effective additives in washing water for decontamination or inhibition of enzymic discoloration
- Gentle spin dry dewatering following washing
- Gentle peeling, cutting, slicing and/or shredding
- Appropriate MAP materials and machinery
- Correct temperatures and handling during chilled storage, distribution and retailing.

Throughout the entire fresh prepared produce chain, four critical factors combine to primarily determine product quality, safety and achievable shelf-life. These critical factors are high quality raw materials, chilled temperature control, appropriate packaging and safe handling (IFPA, 1996). Implementation of a comprehensive quality management system, which incorporates appropriate Good Manufacturing Practices (GMPs), Codes of Practices (CoPs) and Hazard Analysis Critical

Control Points (HACCP) procedures, is highly recommended. Extensive reference to currently available GMP, CoP and HACCP documents should be made before setting up a quality management system (Day, 1992; IFPA, 1993; IFPA, 1996; CFA, 1997; Leaper, 1997; WHO, 1998; Chambers, 1999; CHGL, 1999; IFPA, 1999; Knight, 1999; Seymour, 1999; IFPA, 2000). Brief details from the above GMP, CoP and HACCP documents are incorporated below along with text that concentrates specifically on providing overall guidelines for high O_2 MAP and non-sulphite dipping of fresh prepared produce.

3.1.1 Produce raw materials

Produce raw materials are grown in agricultural soils and so are inevitably covered with soil microorganisms, most of which are completely harmless. Large numbers of microorganisms on produce do not necessarily impair quality. In fact, the initial integral condition of produce items will usually have more effect on achievable shelf-life than intrinsic microbial counts. However, pathogenic bacteria such as *Listeria monocytogenes*, *Salmonella* species, *Shigella* species, *Escherichia coli* O157:H7 and *Clostridium botulinum* could be present and constitute a consumer health hazard (IFPA, 1996). Consequently, it is essential that potential suppliers of produce raw materials use good agricultural practice which includes the adoption of management practices that minimise the risks of water, soil and air contamination (Chambers, 1999). Such good agricultural practice also covers important produce quality and safety related issues such as choice of land, land use history, use of fertilisers and pesticides, irrigation, disease control treatments, harvesting and handling (IFPA, 1996; Chambers, 1999).

Fresh prepared produce manufacturers should ensure that any produce raw materials taken into their premises are checked to make sure that they are safe and of the desired quality for their intended purpose. Quality assurance on all incoming produce raw materials should be subject to an agreement between the supplier and fresh prepared produce manufacturer. Implementation of a quality management standard such as ISO9000:2000 is recommended as a basis for such an agreement. This should include a HACCP analysis to identify what could go wrong with incoming produce raw materials (Leaper, 1997; CHGL, 1999; Knight, 1999; IFPA, 2000).

Checks and controls undertaken should include vetting and site audits of produce raw material suppliers and conformity to objective and agreed produce specifications (e.g. MAFF, 1996). Records of such checks and controls should be kept and used to monitor the performance of produce raw material suppliers and to assure traceability and due diligence requirements (CHGL, 1999).

3.1.2 Hygiene and handling

Rigorous and systematic control of hygienic practices is essential throughout the entire chain from harvested produce raw materials through to final consumption of fresh prepared produce. In England and Wales, all businesses operating within the fresh produce chain must comply with the specific requirements of the Food Safety (General Food Hygiene) Regulations 1995. Strict conditions of hygiene must be maintained to prevent cross-contamination with pathogenic microorganisms and all known direct and indirect sources of contamination should be monitored and controlled using a HACCP-type approach. Particular reference should be made to the CFA (1997), CHGL (1999) and IFPA (1999) guideline documents which outline in detail good hygienic practices. Further advice and recommendations on food hygiene should be sought from local Environmental Health Officers.

All fresh produce raw materials and fresh prepared produce items need to be stored and handled appropriately to avoid unnecessary damage and contamination. Any produce that fails to meet agreed specifications (e.g. MAFF, 1996) should be identified and kept separate from good stock to prevent any danger of cross-contamination or accidental use. Produce raw materials should be identified and kept separate from in-process and packaged fresh prepared produce. Stock rotation is important and storage areas need to be designed (e.g. with racking, bays or area markings) so that stock is used in the correct order (CHGL, 1999).

Throughout all the minimal processing operations involved in the production of fresh prepared produce items, it is of importance to protect the produce from damage caused by poor handling or machinery, foreign body contamination or pest infestation. For example, damage and contamination risks from equipment, personnel and pests should be controlled and monitored at all stages using a

HACCP approach. Above all, good hygiene throughout should be practised and all food handlers should be supervised and instructed and/or trained in food hygiene matters commensurate with their work activities (CHGL, 1999).

3.1.3 Temperature control

The importance of proper temperature control to retard the quality deterioration and assure the microbial safety of fresh prepared produce cannot be overemphasised. In England and Wales, fresh prepared produce items are subject to the specific requirements of the Food Safety (Temperature Control) Regulations 1995. These regulations require foods which are likely to support the growth of pathogenic microorganisms and/or the formation of toxins, to be held at or below 8°C (CHGL, 1999).

It should be noted that whole produce items are not subject to the above temperature control regulations, even though produce is often chilled to maintain its quality. However, certain whole produce items, most notably those from tropical regions, are susceptible to chilling injury by exposure to temperatures below 8°C. Chilling injury may manifest itself as surface pitting or tissue browning, and hence susceptible whole produce items (e.g. bananas, papayas, avocados, melons, tomatoes, cucumbers, pineapples and peppers) should be stored at recommended temperatures and humidities (Holdsworth, 1983). Even though certain whole produce items are susceptible to chilling injury, this injury does not manifest itself when the produce is prepared, since the surface skin is usually removed during preparation (Day, 1994). Hence, fresh prepared produce items need to be stored at or below 8°C.

Chilled stores, distribution vehicles and retail display cabinets should have sufficient refrigeration capacity to maintain fresh prepared produce items at or below 8°C. This refrigeration capacity should be able to cope with conditions of high ambient temperatures and frequent door opening where applicable (Day, 1992). It is good practice to arrange transfers to and from chilled stores, vehicles and cabinets so that exposure times at ambient temperatures are minimised (CHGL, 1999).

Chilled stores, distribution vehicles and retail display cabinets are only designed to maintain the temperature of foods that have already been chilled and cannot be relied upon to reduce the temperature of inadequately chilled foods. The proper refrigerated temperature of each batch of packaged fresh prepared produce should be assured prior to chilled storage, distribution and retail display (Day, 1992). Packaged fresh prepared produce temperatures can be checked by carefully placing an appropriately calibrated thermometer probe (not mercury-in-glass) between two packages which are then pressed together (IFPA, 1999). Careful monitoring of temperature during storage, distribution and retail display is critical and should form part of a quality assurance program based on HACCP principles. Such monitoring ensures that refrigeration equipment is functioning properly. If monitored temperatures fall outside specified ranges, then corrective action should be taken immediately (Day, 1992).

Chilled stores, distribution vehicles and retail display cabinets for fresh prepared produce should be operated at temperatures as low as possible without the possibility of freezing the produce items (ideally, 0-3°C). Good chilled temperature control is the single most important factor for maintaining the quality and assuring the microbial safety of fresh prepared produce, and hence every possible effort should be devoted to assuring the integrity of the chill chain (Anon., 2000). The important influence of storage temperature on the achievable shelf-life of high O_2 MA packed fresh prepared produce is explained in more detail in section 3.3.5.

3.1.4 Preparation procedures

3.1.4.1 Trimming and peeling

Most leafy salad vegetables and some fruits require trimming removal of unwanted parts (e.g. outer leaves, tops, stalks, calyx and core) before further processing. Such initial preparation procedures lead to product losses of 20-70% of the weight of the incoming produce raw materials. These manual procedures are labour intensive and are critical regarding the productivity yields, economics, compositional proportions of produce mixtures and quality of fresh prepared produce items.

The use of high quality produce raw materials (see section 3.1.1) and hygienic storage and handling conditions (see sections 3.1.2 and 3.1.3) greatly facilitate trimming procedures and are of vital importance for the production of high quality fresh prepared produce items (Ahvenainen, 1996). Trimming procedures are best carried out in segregated, hygienic, temperature controlled (10-15°C) factory areas. Unwanted produce parts should be immediately separated from the edible parts and disposed of outside the factory in designated areas. The remaining edible produce parts should be quickly (<10 minutes) conveyed to other segregated, hygienic, temperature controlled factory areas where further processing operations are carried out (see sections 3.1.4.2 and 3.1.4.3).

Most root vegetables (e.g. potatoes, carrots and swede) and some fruit (e.g. apples, oranges, pears and melons) require peeling to remove their outer skins prior to further processing. Such produce items need to be washed carefully in clean potable water and/or disinfected (see section 3.1.4.3) before peeling and any damaged or spoiled parts should be removed.

The method of peeling for the above produce items is one of the most important factors affecting subsequent quality. Industrially, peeling is normally accomplished mechanically (e.g. using rotating carborundum drums or knife blades), but for potatoes in particular, steam peeling and caustic chemical peeling methods are often used. In order to minimise damage to produce cell structures, peeling should be as gentle as possible. Manual peeling with very sharp knives has been shown to cause the least damage but is not compatible with high volume production. Carborundum, steam and caustic chemical peeling methods all cause considerable damage to surface cell structures with consequent enhanced respiration rates, enzymic discolorations, textural deteriorations and microbial growth (Ahvenainen, 1996).

Hence, mechanical knife peeling systems that resemble manual peeling with very sharp knives are recommended. Alternatively, two-stage peeling using a mild carborundum treatment followed by mechanical knife peeling could be utilised to minimise surface cell structural damage. In addition, all peeling machinery needs to be subject to regular cleaning and disinfection routines so as to avoid build-up of organic residues that could encourage microbial growth (Heard, 2000). Also, after any peeling treatment, individual produce items may still need to be hand trimmed before further processing.

3.1.4.2 Cutting and slicing

Many different cutting machines are commercially available and they are capable of grating, chopping, shredding, slicing, chipping or cutting fresh produce into pieces of various shapes and sizes. These machines use cutting devices which are placed perpendicularly to the stream of produce items where sharp blades or discs rotate at high speed. Alternatively, cutting machines are available which use turbines and centrifugal force to propel produce items onto static sharp blades. It should be noted that certain whole fruits (e.g. melons, pineapple, mango and papaya) need to be washed and disinfected (see section 3.1.4.3) before cutting and slicing so as to minimise cross-contamination of the edible internal tissues during cutting operations.

Cutting and slicing with dull blades impairs the quality retention of fresh prepared produce (Ahvenainen, 1996). Therefore, similar to peeling, cutting and slicing with very sharp blades is highly recommended so as to minimise produce cell structural damage. It is important that cutting blades are maintained and sharpened at regular intervals so as to limit damage to produce cell structures. Also, all cutting machinery must be installed securely and safely guarded. In addition, cutting machinery needs to be cleaned and disinfected at regular intervals so as to avoid build-up of organic residues (Heard, 2000).

3.1.4.3 Washing and decontamination

Following, or prior to, trimming, cutting and/or slicing, fresh whole and prepared produce items are subjected to washing and decontamination procedures which are usually carried out in segregated, hygienic, temperature controlled (c. 10°C) factory areas. The objectives of these procedures are to remove unwanted dirt, soil, insects and foreign matter; to reduce spoilage and pathogenic microbial numbers; and to retard enzymic discoloration reactions. However, currently there are no standardised washing and decontamination procedures which are used in the fresh prepared produce industry (Seymour, 1999).

Washing generally involves the immersion of fresh produce in agitated, chilled (5-15°C) potable water whereas decontamination refers to washing in agitated chilled potable water that contains a disinfectant (e.g. chlorine). It should be noted that certain soft fruit (e.g. raspberries, grapes and strawberries) and mushrooms have a diminished shelf-life after immersion in water, so these types of fresh produce items are preferably exposed to gentle sprays and mists for decontamination purposes (DeRoever, 1998).

Chlorine is the most widely used disinfectant for fresh produce because it is relatively inexpensive and, if used correctly, is very effective (Elphick, 1998). The types of chlorine compounds used in wash, spray and flume waters include gaseous chlorine, sodium hypochlorite and calcium hypochlorite. In aqueous solution, hypochlorous acid (HOCl), i.e. "free chlorine", is formed and it is only the HOCl content which is effective as a microbiocide. The dissociation of HOCl is dependent on pH, and hence, in order to maximise the proportion of HOCl, the pH should be maintained in a practical range (typically, 6.8-7.0) to ensure optimal disinfection without causing undue corrosion of metal containers or processing equipment (WHO, 1998). A chlorine concentration of 100ppm is typically used for produce disinfection purposes but it is imperative to monitor pH and free chlorine levels in wash waters in order to maintain effective disinfection. Monitoring and chemical dosing will be required at regular intervals since the organic load and fresh potable water will continually dilute the levels of HOCl in the wash water (Seymour, 1999).

If fresh produce is particularly dirty, it is advisable to carry out a pre-wash in potable water prior to a decontamination wash. Chlorine has an affinity for organic matter and the more organic matter (e.g. soil, leaves and insects) in the washing tank, the more chlorine is lost. Hence, a pre-wash stage will reduce the organic loading on fresh produce and optimise the available free chlorine capacity during a decontamination wash stage.

Although chlorine is the most commonly used disinfectant for decontamination washing of fresh produce, there are increasing concerns over the presence and safety of disinfectant by-products (DBPs) that are found. Because of these concerns, alternative disinfectants such as chlorine dioxide, ozone, organic acids, hydrogen peroxide, trisodium phosphate, sucrose esters and bioflavonoids are being commercially used or explored because they produce fewer DBPs. If chlorine decontamination washing is used, then a final rinse in potable water will help to reduce the levels of residual chlorine and DBPs on fresh prepared produce. It should be noted that chlorine is not permitted as a wash water disinfectant in certain countries (e.g. Germany) and for certain applications (e.g. organic produce) and so many fresh prepared produce companies use approved alternative disinfectants or potable water only for washing. Nevertheless, vigorous washing in potable water has been demonstrated to reduce the number of microorganisms on fresh produce by 10 to 100-fold and can be as effective as decontamination wash treatments using 100-200 ppm chlorine (WHO, 1998). Also, physical decontamination techniques such as ultrasonics, ultraviolet light, pulsed electrical fields and photodynamics have been advocated although their efficacies at reducing spoilage and pathogenic microorganism numbers and the practical consequences of their potential use need to be investigated further (Wiley, 1994; Simons and Sanguansri, 1997; Seymour, 1999).

During washing and decontamination procedures, it is highly recommended to agitate the wash water, which can be done mechanically or by air and/or water jets (Simons and Sanguansri, 1997). Suspended solids and debris should be removed by filters. The recommended quantity of water that should be used is 5-10 litres/kg of produce before peeling and/or cutting (Wiley, 1994) and 3 litres/kg after peeling and/or cutting (Ahvenainen, 1996). Contact times between the produce and the wash water are usually less than 5 minutes throughout all the washing steps (pre-wash, potable water, disinfection and/or rinse) and can be varied by aeration values, altering conveyer speeds, extending the length of the flume or changing the number of washing steps (Simons and Sanguansri, 1997). It must be ensured that each stage of the washing process is sufficient to optimise the removal of dirt, soil, insects, foreign matter and microbial contamination (Seymour, 1999).

Further information and guidance on washing and decontamination of fresh produce can be obtained from the reviews of Wiley, 1994; Hurst, 1995; Ahvenainen, 1996; Simons and Sanguansri, 1997; DeRoever, 1998; Elphick, 1998; WHO, 1998; and Seymour, 1999.

3.1.4.4 Dewatering

After washing, decontamination and/or rinsing, it is important to gently remove water and any DBPs in order to maintain fresh prepared produce quality and safety. Dewatering systems include centrifugal spin dryers, vibrating racks, rotating conveyors, hydro-sieves and spinless drying tunnels (Simons and Sanguansri, 1997). It is recommended that wash water is gently removed since excessive dewatering damages produce and can lead to rapid quality deterioration (Wiley, 1994). Therefore, if spin dryers are used, centrifugation speed and time should be chosen and controlled carefully to limit produce damage. Simple draining is recommended for delicate fresh prepared produce items such as soft fruits. After dewatering, fresh prepared produce items should be quickly and hygienically conveyed to the temperature controlled weighing and packaging areas of the factory.

3.1.5 Quality assurance tests

Quality assurance tests and procedures should be devised as part of a comprehensive quality management system based on HACCP principles. This will require input from technically competent personnel capable of identifying the critical control points in the system, establishing control options, monitoring procedures for these points and enforcing compliance (Leaper, 1997; CHGL, 1999; Knight, 1999; IFPA, 2000).

Appropriate quality assurance tests include compliance with agreed fresh whole and prepared produce specifications (e.g. see Appendix III), microbiological testing (e.g. for total viable counts, total coliforms, Enterobacteriaceae, yeasts, moulds, *Escherichia coli*, *Listeria monocytogenes* and *Salmonella* species), temperature monitoring, HOCl and pH monitoring of decontamination wash waters, and gas analysis and seal integrity testing of packaged fresh prepared produce. Details on these quality assurance tests and acceptable tolerance limits, where appropriate, are available from Day, 1992; Nguyen-the and Carlin, 1994; Betts, 1996; IFPA, 1996; MAFF, 1996; CFA, 1997; DeRoever, 1998; WHO, 1998; IFPA, 1999; Seymour, 1999; and Heard, 2000.

3.1.6 Labelling

In the UK, all packaged fresh prepared produce which is ready for delivery to the ultimate consumer must comply with the Food Labelling Regulations 1996, as amended. These regulations require all packaged food to be marked or labelled with:

- the name of the food;
- a list of ingredients, in descending order by weight, and including any additives, as permitted under the Miscellaneous Additives in Food Regulations 1995;
- the weight of certain ingredients;
- the net weight;
- the appropriate durability indication (e.g. "use by date");
- any special storage conditions (e.g. "keep refrigerated" or "store below 5 °C");
- any special conditions of use (e.g. "consume within x days of opening");
- any cooking or other instructions for use, if appropriate;
- the name and address of the manufacturer, packer or seller established within the EU; and
- the place of origin of the food, where its omission could mislead the consumer as to the food's true origin.

The Food Labelling Regulations 1996, as amended, also require foods that have had their durability extended by means of being packaged in permitted packaging gases (i.e. O_2, CO_2, N_2, Ar, N_2O and helium) to be labelled with the indication "packaged in a protective atmosphere". Extensive further guidance on complying with the requirements of food labelling legislation is available from CCFRA, 1999a. It should also be noted that additional labelling requirements may be imposed by individual national product marketing standards.

3.2 Non-sulphite dipping

As previously explained in section 2.3, the increased regulatory restrictions on the use of sulphites have created an urgent need for safe, practical and functional alternatives for inhibiting enzymic discoloration of fresh prepared produce (Anon., 1991; Ahvenainen, 1996; Laurila *et al*, 1998;

Duncan, 1999). Numerous chemical non-sulphite dip formulations (typically containing mixtures of ascorbic acid, citric acid, malic acid and/or sodium chloride) are commercially available or can be prepared in-house. Whichever non-sulphite dip formulation is used, successful application depends on several important factors which need to be optimised. The following set of general advice and guidance is intended to maximise the performance of non-sulphite dips.

3.2.1 Produce raw materials

As outlined in sections 3.1.1, 3.1.2 and 3.1.3, fresh prepared produce manufacturers must ensure that produce raw materials taken into their premises are safe, of the desired quality, and stored and handled appropriately to avoid unnecessary damage and contamination. Specifically in relation to produce raw materials that are subsequently prepared and non-sulphite dipped, it is an unrealistic assumption to expect chemical non-sulphite dips to overcome quality problems caused by using substandard raw materials which have heavy bruising and/or other major blemishes. Consequently, it is recommended that produce raw materials conform to objective and agreed specifications and are stored and handled as gently as possible so as to minimise bruising. Also, it is recommended that suitable produce raw material cultivars, that have a low tendency to discolour after subsequent preparation, are preferentially selected so that non-sulphite dipping treatments can be minimised (Sapers, 1993). In addition, it is advisable that selected produce raw materials should be of appropriate maturity and firmness so that they can withstand the rigours of subsequent preparation procedures whilst being sufficiently ripe to be of good eating quality.

3.2.2 Pre-dipping preparation treatments

Pre-dipping preparation treatments, such as trimming, peeling, cutting, slicing, washing and decontamination, are no different from those used for fresh produce items that are not subsequently non-sulphite dipped. Hence, it is recommended that the advice and guidance given in sections 3.1.4.1, 3.1.4.2 and 3.1.4.3 be followed, prior to chemical non-sulphite dipping procedures.

If chlorine or other oxidising decontamination agents are used, then it is recommended that fresh prepared produce items be subjected to a final rinse in potable water prior to non-sulphite dipping. This final rinse step will help to reduce the levels of residual oxidising agents, which if too high can substantially counteract the antioxidant properties of the constituents of non-sulphite dips. As a guide, 0.5 litres of rinse water should be applied for every kg of fresh prepared produce.

3.2.3 Dipping procedures

It is recommended that non-sulphite dipping procedures should be applied as soon as possible after the fresh produce has been prepared. Enzymic discolorations usually proceed very rapidly after peeling, cutting and/or slicing, and in the case of certain potato and apple cultivars, discolorations are visible within minutes of preparation. Hence, when immediate immersion in a non-sulphite dip is not possible, fresh prepared produce items should be immersed temporarily in chilled potable water to inhibit enzymic discoloration. The drawback of temporary water immersion for more than a few minutes is that fresh prepared produce items can absorb extensive amounts of water which can lead to soft waterlogged textures, translucent appearances and faster deterioration rates.

Regarding chemical non-sulphite dip formulations and dipping protocols, it should be appreciated that the variables of dip concentration, dipping time and temperature need to be optimised for each fresh prepared produce application. Nevertheless, the following general guidance should be adhered to:

- The constituents of non-sulphite dip formulations must be safe and approved for food use.
- Non-sulphite dipping should not cause any detrimental effects to the flavour, odour, texture and nutritional quality of fresh prepared produce.
- The temperature of the non-sulphite dipping solution should be 0-5°C, since non-sulphite dip constituents are absorbed into plant tissues at a more rapid rate when the dipping solution is at a colder temperature than the temperature of the fresh prepared produce item, prior to dipping.
- Non-sulphite dip solution concentrations are typically in the range of 1-3% w/v. For example, a common non-sulphite dip formulation is diluted in chilled potable water to give a final concentration of 2% w/v ascorbic acid, 1% w/v citric acid and 1% w/v sodium chloride.

- Non-sulphite dipping times are usually in the range of 2-5 minutes. Very short dipping times (<1 minute) are generally not long enough to permit sufficient adsorption of the non-sulphite constituents into cut plant tissues. Conversely, very long dipping times (>5 minutes) are generally not necessary and can lead to excessive water absorption.
- Non-sulphite dipping solutions should be freshly made up and replaced at appropriate intervals depending on the quantity of fresh prepared produce that needs to be dipped.
- The amount of non-sulphite dipping solution required is usually in the range of 1 litre per 10-15 kg of fresh prepared produce to be dipped.

3.2.4 Post-dipping treatments

- A final potable water rinse-off step is typically required after chemical non-sulphite dipping of fresh prepared produce items. This final rinse prevents the persistence of any acidic flavour taints due to the organic acids used in all non-sulphite dip formulations. Typically, 250-500 ml of chilled potable water per kg of fresh prepared produce is used, depending upon variables such as the dip concentration, dipping time and the specific produce application. Another important reason for the final rinse is that the prior non-sulphite dipping treatment can be classified as a processing aid and hence the constituents of non-sulphite dip formulations do not need to be labelled as additives (see section 3.1.6).
- Post non-sulphite dipping and rinsing treatments, such as dewatering, packaging and temperature control, are no different from those used for fresh prepared produce items that are not non-sulphite dipped. Hence, it is recommended that the advice and guidance outlined in sections 3.1.4.4, 3.3.4 and 3.3.5, be followed.

3.3 High O_2 MAP

It should be appreciated that the potential applications of high O_2 MAP technology are a recent innovation and new knowledge will evolve in the future. Hence, the following guidance provided only reflects the current status of available knowledge and experience of high O_2 MAP for fresh prepared produce. Potential applications of high O_2 MAP to chilled combination food items (e.g. chilled ready meals, pizzas, kebabs and sandwiches) are currently the subject of on-going research (Day *et al*, 1999; Day, 2000b) and are outside the scope of this guidelines document. Nevertheless, knowledge gaps and possible research directions are highlighted, in order to assist researchers in the future (see section 5).

3.3.1 Safety

As outlined in section 4.3.1, a specific guidelines document on "The safe application of oxygen enriched atmospheres when packaging food" has been recently published and is publicly available (BCGA, 1998). This document contains clear and concise advice and recommendations on how to safely control the hazards of utilising O_2-rich gas mixtures for the MAP of food. Table 2 (section 4.3.1) lists the contents of this safety guidelines document which is available from CCFRA and the BCGA.

Food companies and related industries (e.g. gas companies and MAP machinery manufacturers) are strongly encouraged to purchase this safety guidelines document and to closely follow the advice and recommendations given before undertaking any pre-commercial trials using high O_2 MAP. Further advice and help on the safety aspects of high O_2 MAP can be sought from qualified gas safety engineers and the BCGA.

3.3.2 Optimal gas levels

Based on CCFRA's practical experimental trials (see section 4.3.3), the recommended optimal headspace gas levels immediately after fresh prepared produce package sealing are:

80-95% O_2 / 5-20% N_2

After package sealing, headspace O_2 levels will decline whereas CO_2 levels will increase, during chilled storage, due to the intrinsic respiratory nature of fresh prepared produce (see section 1.4). As previously explained, the levels of O_2 and CO_2 established within hermetically sealed packs of produce during chilled storage are influenced by numerous variables, i.e. the intrinsic produce respiration rate (which itself is affected by temperature; atmospheric composition; produce type, variety, cultivar and maturity; and severity of preparation); packaging film gas permeability; pack volume, surface area and fill weight; produce volume/gas volume ratio and degree of illumination (Kader et al, 1989; Day, 1994; O'Beirne, 1999).

To maximise the benefits of high O_2 MAP, it is desirable to maintain headspace levels of O_2 > 40% and CO_2 in the range of 10-25% during the chilled shelf-life of the product. This can be achieved by lowering the temperature of storage (see sections 3.1.3 and 3.3.5); by selecting produce having a lower intrinsic respiration rate (see Appendix I); by minimising cut surface tissue damage (see section 3.1.4); by reducing the produce volume/gas volume ratio by either decreasing the pack fill weight or increasing the pack headspace volume (see section 3.3.3); by using a packaging film which can maintain high levels of O_2 whilst selectively allowing excess CO_2 to escape (see section 3.3.4); or by incorporating an innovative active packaging sachet that can scavenge excess CO_2 and emit an equal volume of O_2 (see section 5 and McGrath, 2000).

Also, in order to maintain levels of O_2 > 40% and CO_2 in the range 10-25% during the chilled shelf-life of the product, it is desirable to introduce the highest level of O_2 (balance, N_2) possible just prior to fresh prepared produce package sealing. Generally, it is not necessary to introduce any CO_2 in the initial gas mixture since levels of CO_2 will build up rapidly within sealed packages during chilled storage.

However, for fresh prepared produce items that have low intrinsic respiration rates, are packaged in a format with a low produce volume/gas volume ratio (see sections 3.3.3 and 3.3.4), are stored at low chilled temperatures (see section 3.3.5), or have an O_2 emitter/CO_2 scavenger sachet incorporated into the sealed package (see section 5), then the incorporation of 5-10% CO_2 into the initial gas mixture may be desirable (see section 1.4). Based on the results of controlled atmosphere storage experiments, the most effective high O_2 gas mixtures were found to be 80-85% O_2/15-20% CO_2; these had the most noticeable sensory quality and antimicrobial benefits on a range of fresh prepared produce items (see sections 4.3.3, 4.3.4 and 4.3.5).

The type of MAP machinery used will greatly influence the maximum achievable O_2 level that can be introduced just prior to fresh prepared produce package sealing. As mentioned previously (see section 1.4 and BCGA, 1998), most light prepared salad items are commercially MA packed on VFFS and HFFS machines (Hartley, 2000). These machines use a gas flushing or air dilution technique to introduce gas in MA pillow-packs just prior to sealing. Since these machines do not use an evacuation step, then c.80% O_2 would be the highest practical level that could be achieved within sealed fresh prepared produce packs by initially flushing with 100% O_2. Higher levels of in-pack O_2 could be achieved by substantially increasing the flow rate of O_2 through the gas flushing lance of "flow pack" machines, but this is not recommended for economic and safety reasons (see BCGA, 1998).

In contrast to VFFS and HFFS machines, TFFS, PTLF, VC and ST machines (see section 1.4 and BCGA, 1998) use a compensated vacuum technique to evacuate air and then introduce gas into tray and lidding film and/or flexible MA packs. Since these machines use an evacuation step prior to gas (i.e. 100% O_2) introduction, much higher levels of headspace O_2 (85-95%) can be achieved within such sealed fresh prepared produce packs. Also, all compensated vacuum machines (except VC machines) are intrinsically safer for high O_2 MAP applications, compared with VFFS and HFFS machines, since O_2 is introduced directly into the MA packs after air evacuation and prior to sealing, and consequently O_2 levels in the air surrounding these machines are not enriched (BCGA, 1998).

3.3.3 Produce volume/gas volume ratio

In order to maintain headspace O_2 levels > 40% and CO_2 levels in the range 10-25% during the chilled shelf-life of the product, it is desirable to minimise the produce volume/gas volume ratio of fresh prepared produce MA packs. This can be achieved by either decreasing the pack fill weight or increasing the pack headspace volume. Decreasing the pack fill weight of fresh prepared produce will have the effect of reducing the overall respiratory load or activity within MA packs and hence the rate of O_2 depletion will be reduced. Increasing the pack headspace volume will have the effect of increasing the reservoir of O_2 for respiratory purposes and hence the rate of O_2 depletion will also be reduced. Consequently, low produce volume/gas volume ratios are conducive to maintaining headspace O_2 levels >40% and CO_2 levels within the range 10-25%.

The important influence of the produce volume/gas volume ratio, in addition to the intrinsic produce respiration rate and packaging film gas permeability (see section 3.3.4), is well illustrated by the results from CCFRA's bulk iceberg lettuce trial #17 (see section 4.3.3 and Appendix IV, Table 5). Depletion of O_2 and elevation of CO_2 levels within the high O_2 MA bulk packs of this trial were very rapid because these packs contained 2 kg of fresh prepared iceberg lettuce as opposed to only 200g for retail MA packs. Consequently, the produce volume/gas volume ratio and overall respiratory load were much higher in these MA bulk packs compared with MA retail packs. Also, the iceberg lettuce used for this bulk pack trial was shredded (10 mm cut) and hence had a much higher intrinsic respiration rate compared with retail salad cut (40-70 mm) iceberg lettuce. In addition, the thicker (60μm compared with 30μm for retail) and less permeable bulk OPP/LDPE bags exacerbated the depletion of O_2 and elevation of CO_2. Hence, it was not surprising that the achievable shelf-life at 8°C for high O_2 MA bulk packed fresh shredded iceberg lettuce was found to be only 2 days, even though the shelf-life of equivalent low O_2 MA bulk packed iceberg lettuce was even shorter (see Appendix IV, Table 5).

It should be appreciated that there are practical and commercial limits to the reduction of produce volume/gas volume ratios for fresh prepared produce MA packs. Obviously, retail consumers will not readily accept MA packs of fresh prepared produce which appear underfilled with too much headspace gas. Therefore, it is recommended that potential users of high O_2 MAP technology should carry out pre-commercial trials with fresh prepared produce packs having different but practical produce volume/gas volume ratios.

3.3.4 Packaging materials

Based on the results of CCFRA's practical experimental trials (see section 4.3.3), the recommended packaging material for high O_2 MA retail packs of fresh prepared produce is:

30 µm orientated polypropylene (OPP) with anti-fog coating

It should be noted that initial experimental trials carried out at CCFRA on high O_2 MAP of fresh prepared produce used an O_2 barrier film, i.e. 30µm polyvinylidene chloride (PVDC) coated OPP with anti-fog coating, because it was considered at the time to be important to maintain the highest levels of O_2 within high O_2 MA packs. However, extensive experimental trials on high O_2 MAP of fresh prepared iceberg lettuce using 30µm PVDC coated OPP film (see section 4.3.3 and Appendix IV, Table I) clearly demonstrated that excess and potentially damaging levels of CO_2 (30-40%) could be generated within such O_2 barrier film packs, particularly at higher chilled storage temperatures (i.e. 6-8 °C). Consequently, 30µm OPP film was used for subsequent high O_2 MAP experimental trials, instead of 30µm PVDC coated OPP film, and for the majority of fresh prepared produce items, was found to have sufficient O_2 barrier properties to maintain high in-pack O_2 levels (>40%) but also be sufficiently permeable to ensure that in-pack CO_2 levels did not rise above 25%, after 7-10 days storage at 5-8 °C (see Appendix IV, Tables 1-5).

It should be appreciated that other packaging materials, apart from 30µm OPP, may be suitable for high O_2 MAP of fresh prepared produce (Air Products, 1995; Day and Wiktorowicz, 1999). For example, laminations or extrusions of OPP with low density polyethylene (LDPE), ethylene-vinyl acetate (EVA) and polyvinyl chloride (PVC) or other medium to very high O_2 permeability films maybe more suitable for high O_2 MAP of fresh prepared produce items that have a higher respiration rate than iceberg lettuce (see Appendices I and II). Also, the produce volume/gas volume ratio of different retail MA pack formats (e.g. pillow packs or tray and lidding film systems), the intrinsic fresh prepared produce respiration rate and chilled temperature of storage will influence the selection of the most suitable packaging film for high O_2 MAP applications (see sections 3.3.2, 3.3.3 and 3.3.5).

It is recommended that potential users of high O_2 MAP for fresh prepared produce should initially carry out pre-commercial shelf-life trials using 30μm OPP with anti-fog coating as the packaging film for flexible pillow packs or as a tray lidding film (see section 4.3.3). Regular gas analyses of the in-pack atmospheres during chilled storage will reveal whether the packaging film is not permeable enough (resulting in build-up of excess levels of CO_2 to >25%) or too permeable (resulting in depletion of O_2 to < 40% and slow build-up of CO_2 to <10%). If the in-pack O_2 levels fall to < 40% and CO_2 levels lie outside the range 10-25% by the end of the chilled shelf-life of the product, then adjustments to the produce volume/gas volume ratio, chilled temperature of storage, seal integrity, pack format and/or gas permeability of the package film will need to be made and further shelf-life trials carried out.

It should also be noted that O_2 barrier films could be used for high O_2 (or low O_2) MAP of fresh prepared produce items if an O_2 emitter/CO_2 scavenger sachet is incorporated into sealed packages (see section 5). Appropriate transparent O_2 barrier films (with anti-fog coatings) include PVDC coated OPP, and coextrusions or laminations containing ethylene-vinyl alcohol (EVOH), polyester (PET), polyamide (nylon) and/or PVDC (Air Products, 1995; Day and Wiktorowicz, 1999).

Whatever packaging material is used for high O_2 MAP applications, all of them must comply with statutory legal requirements. In the UK, these requirements include the Materials and Articles in Contact with Food Regulations 1987, Plastic Materials and Articles in Contact with Food Regulations 1998, Producer Responsibility Obligations (Packaging Waste) Regulations 1997 and Packaging (Essential Requirements) Regulations 1998.

All packaging materials should be purchased to an agreed specification which includes details of technical properties and performance.

Quality assurance on all incoming packaging materials should be subject to an agreement between the packaging supplier and user. Each delivery or batch should be given a reference code to identify it in storage and use, and the documentation should allow any batch of packaged product to be correlated with deliveries of respective packaging materials. All packaging materials should be stored off the floor in separate and dry areas of the factory and should be inspected at regular intervals to ensure that they remain in acceptable condition. Authorised procedures and documentation should be established and followed for the issue of packaging materials from store (Day, 1992). Further advice on the technical requirements, properties, performance and handling of packaging materials should be sought from reliable suppliers.

3.3.5 Temperature control

As detailed in section 3.1.3, the importance of proper temperature control to retard the quality deterioration and assure the microbial safety of fresh prepared produce cannot be overemphasised. For high O_2 MA packed fresh prepared produce, it is recommended that the temperature be maintained below 8 °C, and ideally in the range 0-3°C, throughout the entire chill chain.

The important influences of storage temperature and packaging film gas permeability on the quality of high O_2 MA packed fresh prepared produce can be illustrated by the results from CCFRA's fresh prepared iceberg lettuce trials (Appendix IV, Table I). The results from these trials clearly demonstrated that temperature and packaging film gas permeability are critical factors in determining the development of O_2 and CO_2 levels within high O_2 MA packs, during chilled storage. Higher temperatures of storage correlate to high respiratory rates and hence greater depletion of O_2 and elevation of CO_2 within sealed high O_2 MA barrier (i.e. 30 µm PVCD coated OPP) pillow packs of fresh prepared iceberg lettuce. The most beneficial sensory effects of high O_2 MAP were obtained from trial #1 where the temperature of storage was 3-5°C and the O_2 levels dropped from 70% to 55% and the CO_2 levels reached only 15% after 10 days' storage. In contrast, largely negative sensory effects were obtained from trials #7-12, where an elevated chill temperature of storage regime (8°C) was employed. Under this elevated chilled temperature of storage regime, O_2 levels dropped from 80% to 35-40% whereas CO_2 levels reached 35-40% after 10 days' storage. These high levels of generated CO_2 within the high O_2 MA barrier pillow packs of fresh prepared iceberg lettuce were responsible for the undesirable "CO_2 damage" discoloration observed. As mentioned in section 3.3.4, subsequent high O_2 MAP experimental trials (e.g. iceberg lettuce trial #13) used more permeable OPP film whereby high O_2 (> 40%) levels were generally maintained and CO_2 levels did not rise above 25% after 7-10 days' storage at 5 °C and 8 °C. Under these high O_2 MAP conditions, beneficial sensory effects were observed for the majority of the fresh prepared produce items studied, in comparison with industry standard air and/or low O_2 MAP (see Appendix IV).

3.3.6 Fresh prepared produce applications

As mentioned in section 4.3.3 and detailed in Appendix IV, high O_2 MAP was found to have beneficial effects on the sensory quality of the vast majority of the fresh prepared produce items studied. Under defined storage and packaging conditions (see sections 3.3.2, 3.3.3, 3.3.4 and 3.3.5) and in comparison with industry-standard air packing and/or low O_2 MAP, high O_2 MAP was found to be effective for extending the achievable shelf-lives of retail packs of fresh prepared iceberg lettuce, sliced mushrooms, potatoes, sliced bananas, little gem lettuce, cos lettuce, baby-leaf spinach, raddichio lettuce, lollo rossa lettuce, flat-leaf parsley, cubed swede, coriander, raspberries and strawberries (see Table 5, Appendix IV). In addition, the results from trials carried out prior to September 1997, showed beneficial sensory effects of high O_2 MAP for fresh prepared tomato slices, baton carrots, pineapple cubes, broccoli florets, honeydew melon cubes, sliced mixed peppers and sliced leeks (see Tables 3 and 4, Appendix IV). Also, high O_2 controlled atmospheres were found to extend the shelf-life of table grapes and oranges (see section 4.3.3).

It should be noted that in comparison with industry-standard air and/or low O_2 MAP, high O_2 MAP was not found to have beneficial effects on the sensory quality of retail packs of fresh prepared apple slices, curly parsley, red oak leaf lettuce and Galia melon cubes, and bulk packs of shredded iceberg lettuce (see Tables 3 and 5, Appendix IV). However, it is probable that beneficial effects of high O_2 MAP on the above fresh prepared produce items would have been achieved if the chilled storage temperature, high O_2 gas level, packaging film gas permeability, produce volume/gas volume ratio and/or preparation procedures had been optimised adequately.

Consequently, it is recommended that potential users of high O_2 MAP for specific fresh prepared produce items or combinations carry out pre-commercial optimisation trials by utilising the advice given in sections 3.3.1, 3.3.2, 3.3.3, 3.3.4 and 3.3.5.

4. CCFRA CLUB AND EU FAIR FUNDED RESEARCH

4.1 Objectives

The overall objective of this research was to develop novel MAP applications for fresh prepared, minimally processed produce. The major focus of these novel MAP applications was on high O_2 MAP, followed by Ar MAP and, to a minor extent, N_2O MAP. The overall objective was split into the following key sub-objectives:

- to establish safe commercial applications of novel high O_2, Ar and N_2O MAP for extending the quality shelf-life of a wide selection of prepared produce.
- to investigate the synergy between the use of non-sulphite dips with novel MAP for inhibiting the discoloration of prepared produce.
- to thoroughly investigate the underlying microbial and biochemical spoilage mechanisms which are inhibited under novel MAP conditions.
- to assess the effects of high O_2 MAP on nutritional components (e.g. vitamins C and E, carotenoids, polyphenolics, quinols and polyunsaturated fatty acids) *in vitro* and *in vivo*.
- to compile a comprehensive guidelines document which outlines good manufacturing and handling practices for novel MAP as well as non-sulphite dipping of fresh prepared produce.

4.2 Research Tasks

In order to address the sub-objectives outlined in section 4.1, the research was divided into seven scientific and technical tasks:

1. Investigation of the safe use of high O_2 mixtures with MAP machinery.

2. Assessment of the effectiveness of non-sulphite dips at inhibiting discoloration of prepared produce under various novel MAP conditions.

3. Assessment of the effects of novel MAP on a wide range of selected prepared produce items.

4. Investigation of the effects of high O_2, Ar and N_2O MAP on a wide variety of microorganisms.

5. Investigation of the effects of high O_2 and Ar MAP on various biochemical reactions.

6. Investigation of the effects of high O_2 MAP on nutritional components of selected prepared produce items, *in vitro* and *in vivo*.

7. Compilation of a comprehensive guidelines document for good manufacturing and handling of prepared produce using novel MAP and non-sulphite dipping treatments.

4.3   Summary of major findings

4.3.1   Safety of high O_2 MAP

- Extensive consultations were undertaken with gas safety engineers and manufacturers of commercially available MAP machinery to assure the safety of high O_2 MAP. Several types of MAP machinery were certified as being compatible with high O_2 (>25% O_2) mixtures and were utilised safely and successfully during experimental trial work at CCFRA for the high O_2 MAP of fresh prepared produce items.

- Experimental gas analysis trials were performed at commercial prepared salad factories and the results of these trials demonstrated that the enrichment levels of O_2 in the immediate vicinity of operating vertical form-fill-seal (VFFS) MAP machines were <23.5% O_2. The high O_2 MAP compatibilities of horizontal form-fill-seal (HFFS), thermoform-fill-seal (TFFS), snorkel type (ST), preformed tray and lidding film (PTLF) and vacuum chamber (VC) machinery were also investigated and the results incorporated into a specific safety guidelines document (BCGA, 1998).

- A Safety Sub-Committee Working Group of the CCFRA High O_2 MAP Club was set up and consisted of qualified personnel from gas companies, MAP machinery manufacturers, the UK Health and Safety Executive, CCFRA and the British Compressed Gases Association (BCGA).

After many meetings, consultations, factory visits and reviews of existing safety data during 1995-1997, the Working Group developed a specific guidelines document on "The safe application of oxygen enriched atmospheres when packaging food" (BCGA, 1998). This non-confidential guidelines document provides users of high O_2 MAP with clear and concise advice and recommendations on how to safely control the hazards of utilising O_2-rich gas mixtures for the MAP of food. Table 2 lists the contents of this safety guidelines document which is available from CCFRA and the BCGA (see section 3.3.1).

Table 2: Contents of the guidelines document on the safe use of high O_2 MAP

THE HAZARDS OF OXYGEN
- Properties of oxygen
- Necessary conditions for a fire
- Controlling the fire hazards of oxygen
- Other risks
- Further information on the hazards of oxygen

ADVICE TO PACKAGING MACHINE DESIGNERS AND MANUFACTURERS
- Layout and configuration of oxygen enriched gas pipework and joining techniques
- Selection of materials and components
- Guidelines on cleanliness and workmanship
- Extraction and ventilation requirements
- Warning signs
- Special considerations for particular types of machines
- User instructions

GUIDANCE FOR USERS OF MAP MACHINES WITH OXYGEN ENRICHED ATMOSPHERES
- Hazards of oxygen
- Siting the machine
- Gas supply system
- Oxygen compatibility of packaging materials
- Precautions in use
- Precautions in maintenance
- Precautions in storage and transport of oxygen enriched gas MAP products

REFERENCES
APPENDICES

4.3.2 Non-sulphite dipping and novel MAP

- VTT, Finland, and UL, Ireland, screened numerous non-sulphite organic acids (e.g. ascorbic, citric, lactic, malic, tartaric and succinic acids), antioxidants, chelating agents, edible films, enzymes, salts and PPO inhibitors for their effectiveness at inhibiting enzymic discoloration of fresh prepared potatoes and apples. Substantial evidence was gathered to demonstrate that certain combinations of the above reagents could be functional non-sulphite alternatives for inhibiting enzymic discoloration.

- Several non-sulphite dipping variables (i.e. dip formulation, dip concentration, dip temperature and dip time) were optimised and suitable dipping protocols were recommended for inhibiting the enzymic discoloration of fresh prepared potatoes and apples.

- Sulphite dipped potatoes were generally of similar appearance to equivalent non-sulphite dipped potatoes but suffered from undesirable sulphurous odours and excess case hardening and exudate during chilled storage.

- The important influence of the type of peeling method on potato quality was conclusively demonstrated by experimental trials which showed that gentle knife peeling resulted in far less damage to potato tissues and hence less enzymic discoloration, case hardening and exudate compared with abrasive carborundum peeling.

- Enzymic discolorations of fresh prepared non-sulphite dipped potatoes and apples were generally more effectively inhibited by anaerobic (<2% O_2) MAP combinations of N_2, Ar and CO_2, compared with high O_2 MAP. However, high O_2 MAP was found to have odour and textural benefits for fresh prepared non-sulphite dipped potatoes and apples. Also, high O_2 MA packed non-sulphite dipped fresh prepared potatoes and sliced bananas were found to have longer achievable shelf-lives, in comparison with equivalent low O_2 (8%) MA packed control samples.

- High O_2 MAP was not found to prevent enzymic browning of non-sulphite dipped fresh prepared apple slices. However, the dipped apple slices packed under high O_2 MAP retained their firm texture and did not discolour further upon pack opening. In contrast, anaerobic N_2/Ar MA packed sliced apple controls were prone to textural softening and tended to brown rapidly upon pack opening.

- Ar-containing MAP treatments appeared to have negligible, variable or only minor beneficial effects on the overall sensory quality gradings of both sulphite and non-sulphite dipped fresh prepared potatoes compared with equivalent N_2-containing MAP treatments.

4.3.3 Effects of novel MAP on fresh prepared produce

- CCFRA carried out over 65 practical experimental trials on a wide range of fresh prepared produce items. These experiments ranged in complexity from simple "look-see" comparative screening trials to comprehensive trials which incorporated extensive sensory evaluations, microbiological assessments and gas analyses under many different air, low O_2, high O_2 and Ar-containing MAP conditions, preparation procedures and storage temperatures. Full details from these trials are outlined in 35 comprehensive reports which are confidential to members of the High O_2 MAP Club, Novel Gases MAP Club, the EU Commission and CCFRA's EU partners (see Acknowledgements, p iii-iv).
- In the trials carried out, all produce samples were assigned random three-numbered codes so that the samples were evaluated in an unbiased manner. Duplicate gas analyses were undertaken to assess the development of different atmospheres within sealed packs of the samples during storage under various chilled temperatures. Quintuplicate microbiological assessments for various generic groups of microorganisms were also carried out and the results were statistically analysed (see section 4.3.4).
- Sensory quality of the novel MA packed prepared produce samples was evaluated by examining the overall appearance, odour and texture after appropriate chilled storage dates and comparisons were made against industry standard air packed, low O_2 MAP and/or perforated controls. Sensory quality was assessed by trained sensory panels using free descriptions for the trials undertaken prior to September 1997.
- Sensory quality was assessed by trained sensory panels using specific graded specifications for the trials undertaken after September 1997. An example of a specific graded specification for fresh prepared iceberg lettuce is included in Appendix III. By using such specific graded sensory specifications, the storage time taken for prepared produce samples (packed under different MAP treatments) to fall to grade C or grade 3 on a numerical scale (the unacceptable quality cut-off line) was assessed for the individual quality attributes (i.e. appearance, odour and texture). By

analysing this information (Figure 3) an overall achievable shelf-life (the storage time at a specified temperature for the shelf-life limiting quality attribute to fall to grade C) could be determined for each produce item, packed under each different MAP treatment. To illustrate such sensory analyses, Tables 3 and 4 show the overall achievable shelf-lives obtained for different MAP treatments from the fresh prepared iceberg lettuce trial #14 and sliced mushroom trial #2, respectively.

Figure 3: Schematic representation of the determination of achievable shelf-life for fresh prepared produce

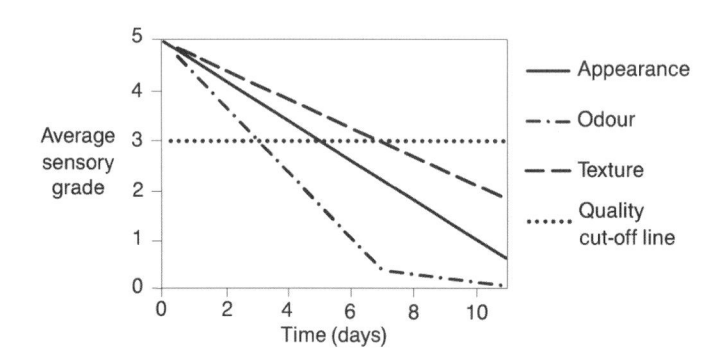

Table 3: Overall achievable shelf-life obtained from fresh prepared iceberg lettuce trial #14

MAP Treatments	Storage days at 8 °C to drop to quality grade C			Shelf-life limiting quality attribute(s)	Overall achievable shelf-life
	Appearance	Odour	Texture		
5% O_2/95% N_2 in OPP film	4	7	4	Appearance / Texture	4 days
5% O_2/10% CO_2/85% N_2 in OPP film	7	7	8	Appearance / Odour	7 days
80% O_2/20% N_2 in OPP film	11	11	11	Appearance / Odour / Texture	11 days

Table 4: Overall achievable self-life obtained from fresh prepared sliced mushroom trial#2

Packaging Treatments	Storage days at 8 °C to drop to quality grade C		Shelf-life limiting quality attribute(s)	Overall achievable shelf-life
	Appearance	Odour		
UNWASHED				
Air (macroperforated film)	5	12	Appearance	5 days
Air (microperforated film)	6	6	Appearance / Odour	6 days
Air (OPP film)	>12	2	Odour	2 days
80% O_2/20% N_2 in OPP film	>12	>12	-----	>12 days
80% O_2/20% Ar in OPP film	>12	9	Odour	9 days
WASHED				
Air (macroperforated film)	2	1	Odour	1 day
Air (microperforated film)	2	4	Appearance	2 days
Air (OPP film)	4	2	Odour	2 days
80% O_2/20% N_2 in OPP film	6	6	Odour / Appearance	6 days
80% O_2/20% Ar in OPP film	6	6	Odour / Appearance	6 days

- Tables 1-5 in Appendix IV summarise the results from CCFRA's fresh prepared produce trials. Of the trials carried out after September 1997 (Table 5, Appendix IV), high O_2 MAP was found to have beneficial effects on the sensory quality of the vast majority of the fresh prepared produce items studied. In comparison with industry standard air packing and/or low O_2 MAP, high O_2 MAP was found to be effective for extending the achievable shelf-lives of fresh prepared iceberg lettuce, sliced mushrooms, potatoes, sliced bananas, little gem lettuce, cos lettuce, baby-leaf spinach, raddichio lettuce, lollo rossa lettuce, flat-leaf parsley, cubed swede, coriander, raspberries and strawberries.

- Additional experimental trials carried out by ATO-DLO, The Netherlands, showed that high O_2 controlled atmospheres inhibited fungal growth and thereby extended the shelf-life of table grapes and oranges, in comparison with air stored control samples.

- Twelve of CCFRA's fresh prepared produce trials that were undertaken from mid-November 1997 to mid-November 1998 incorporated Ar-containing low O_2 and high O_2 MAP treatments. In order to allow for a direct comparison to be made between N_2 and Ar, identical levels of each gas were incorporated into separate low O_2 MAP (i.e. 8%O_2/92%N_2 and 8%O_2/92%Ar) and high O_2 MAP (i.e. 80%O_2/20%N_2 and 80%O_2/20%Ar) treatments. Results achieved from these trials indicated that Ar-containing MAP treatments had negligible or variable effects on generated in-pack MAs and sensory quality gradings of fresh prepared produce in comparison with equivalent N_2-containing MAP treatments. Future investigations may reveal more consistent beneficial effects of Ar-containing MAP treatments but based on these experimental results, high O_2 MAP treatments carry far more promise for extending the quality shelf-life of fresh prepared produce items.

- Additional research carried out by SIK, Sweden, demonstrated that N_2O-containing MAP treatments had negligible effects on the sensory quality of fresh potatoes and onions, in comparison with equivalent N_2-containing MAP treatments.

4.3.4 Effects of novel MAP on microbial growth

- Microbiological assessments carried out by CCFRA on selected prepared produce items (i.e. iceberg lettuce, cos lettuce, potatoes, beansprouts, honeydew melon, sliced mushrooms, broccoli florets and baton carrots) showed that high O_2 (i.e. 70%, 80% and 100%) MAP treatments are capable of inhibiting certain generic groups of microorganisms. These groups included total aerobes (TVC), total anaerobes, yeasts, moulds, *Pseudomonas* species, Enterobacteriaceae and coliforms, and in many statistical analyses, the microbial levels were significantly lower on high O_2 MAP samples than with equivalent low O_2 MAP samples. For example, Figure 4 illustrates the antimicrobial effect of high O_2 MAP after 9 days' storage at 8°C on the growth of *Pseudomonas* species on fresh prepared potatoes. It should be noted that CO_2 is generated within high O_2 MA packs of prepared produce during storage and this CO_2 may be partially responsible for the observed antimicrobial effects. To test this idea, a sliced mushroom trial was carried out where CO_2 scavenger sachets were incorporated into high O_2 MA packs. The results from this trial showed that the combination of high O_2 and generated CO_2 was primarily responsible for the antimicrobial effects observed, rather than high O_2 or generated CO_2 alone.

Figure 4: Growth of *Pseudomonas* species on fresh prepared MA packed potatoes

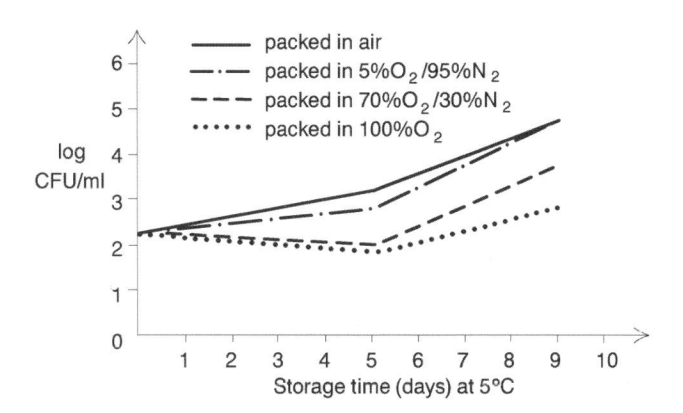

- Results obtained by SIK, Sweden, indicated that high O_2 levels (35-100%) did not inhibit or stimulate the growth of *Pseudomonas fragi, Bacillus cereus* or *Lactobacillus sake,* but did inhibit the growth of *Yersinia enterocolitica,* with higher levels of O_2 causing more pronounced inhibition of this pathogen (Figure 5).

Figure 5: Inhibition of *Yersinia enterocolitica* growth by high O_2 MAs (balance N_2)

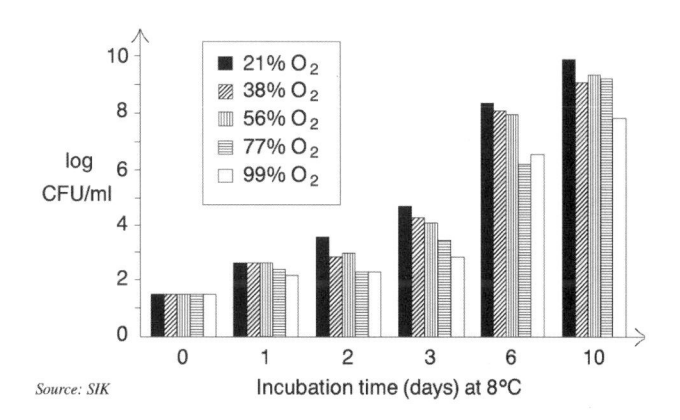

- ATO-DLO, The Netherlands, found that high O_2 levels inhibited the growth rate of *Pseudomonas putida, Lactobacillus sake, Aeromonas hydrophila, Salmonella* Enteritidis, *Pantoea agglomerans* and several lactic acid bacteria and these inhibitions were more pronounced when a combination of high O_2 (90%) and 10% CO_2 was used. In addition, three genera of fungi (*Rhizopus, Botrytis* and *Penicillium*) were found to be significantly inhibited by high O_2 (80-85%) in combination with 15-20% CO_2 and these inhibitions were far more pronounced in comparison with equivalent low O_2 (5-21%) atmospheres (Figure 6).

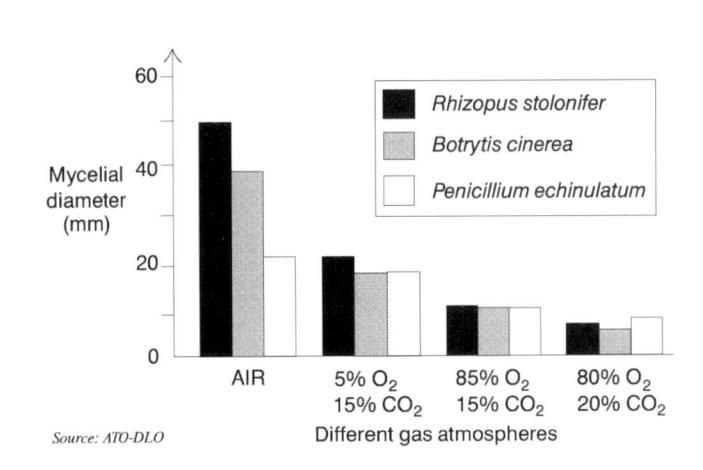

Figure 6: Inhibition of fungal growth by different MAs

- ATO-DLO also carried out high O_2 controlled atmosphere trials on oranges which were infected with *Penicillium digatum* and the results showed that fungal growth (Figure 7) was most effectively inhibited by a combination of high O_2 (80%) and 20% CO_2. These results were similar to the findings of fungal growth inhibition on table grapes stored under 80% O_2/10% CO_2/10% N_2.
- CCFRA investigated the effects of high O_2 MAs in combination with CO_2 on the growth kinetics of a range of spoilage and pathogenic bacteria in a batch fermenter system at 8°C, 20°C and 25°C. The results showed that high O_2 (80%) MAs inhibited the growth of *Aeromonas hydrophila* and this anti-microbial effect was enhanced by a combination of 80% O_2 with 20% CO_2. High O_2 MAs alone did not inhibit or stimulate the growth of *Pseudomonas fragi, Yersinia enterocolitica* and *Listeria monocytogenes*, but the addition of 20% CO_2 inhibited the growth of all these types of bacteria.

Figure 7: Inhibition of fungal growth on *Penicillium digitatum* infected oranges under different MAs

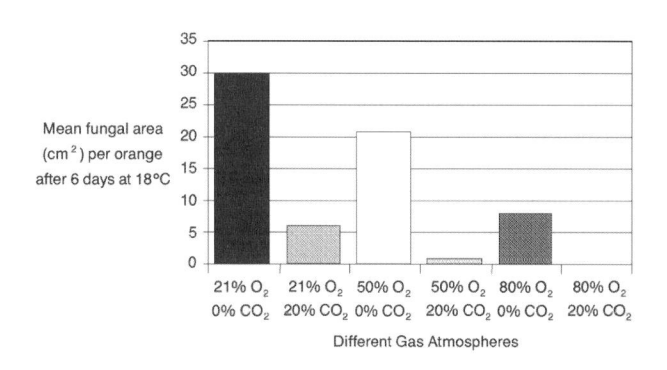

[Bar chart showing Mean fungal area (cm²) per orange after 6 days at 18°C for different gas atmospheres:
- 21% O₂, 0% CO₂: ~30
- 21% O₂, 20% CO₂: ~6
- 50% O₂, 0% CO₂: ~21
- 50% O₂, 20% CO₂: ~1
- 80% O₂, 0% CO₂: ~8
- 80% O₂, 20% CO₂: ~0]

Source: ATO - DLO

- Experimental trials carried out by CCFRA and SIK showed that Ar-containing and N₂O-containing MAs did not have any additional antimicrobial effects on a wide range of spoilage and pathogenic microorganisms, when compared with equivalent N₂-containing MAs.

4.3.5 Effects of novel MAP on biochemical reactions

- Research undertaken by UL, Ireland, showed that respiration rates of selected fresh prepared produce items (carrot discs, apple slices and shredded lettuce) were not found to be significantly affected by high O₂ (80%) or high Ar (40%) MAs, in comparison with air at 3°C or 8°C. As expected, respiration rates at 3°C were significantly lower than those at 8°C. The respiration rates of fresh prepared carrot discs and shredded lettuce at 8°C were significantly reduced when 10% CO₂ was combined with 80% O₂, but a similar reduction in respiration rates was measured when 10% CO₂ was combined with air. ATO-DLO, The Netherlands, found that under very low O₂ (<2%) MAs, the respiration rates of table grapes and oranges increased due to anaerobic fermentation processes.

- Beneficial effects of high O₂ and high Ar MAP were found by UL, Ireland, on the retention of ascorbic acid, indicators of lipid oxidation, sensory quality and the inhibition of enzymic discoloration of fresh prepared shredded lettuce, at 3°C and 8°C (Barry-Ryan and O'Beirne, 1999).

- High O_2 and high Ar MAP were not found to adversely affect the cell permeability, tissue exudate, lipoxygenase activity or tissue pH of fresh prepared carrot discs, at 3°C and 8°C.
- High O_2 and high Ar MAP alone did not prevent enzymic browning of non-sulphite dipped fresh prepared apple slices, but no further browning took place after pack opening.
- High O_2 MAP increased the membrane damage of fresh prepared apple slices, whereas high Ar MAP decreased membrane damage. However, apple slices stored under O_2-free MAs suffered the most membrane damage, which adversely affected tissue integrity, cell leakage and texture.
- Mushroom polyphenol oxidase (PPO) activity was totally inhibited under anaerobic 100% N_2 and 100% Ar MAP conditions. Mushroom PPO activity was found to be substantially inhibited under 90% Ar/10% O_2 when compared with 90% N_2/10% O_2.
- Conflicting results were obtained regarding the inhibition of mushroom PPO by high O_2 MAs. Initial experimental results showed that mushroom PPO activity was significantly inhibited under 80% O_2/20% N_2 when compared with air storage. However, latter experimental results showed no significant inhibition of mushroom PPO activity under 80% O_2/20% N_2. However, the incorporation of 20% CO_2 into high O_2 MAs may possibly inhibit mushroom PPO activity as well as the activity of other prepared produce PPOs (Sapers, 1993).
- ATO-DLO, The Netherlands, found that the peroxidase activity of *Botrytis cinerea* was increased under high O_2 MAs, but the addition of 10% CO_2 substantially reduced this activity.

4.3.6 Effects of high O_2 on nutritional components

- INN, Italy, carried out analyses of sample extracts from fresh prepared kale, lettuce, onions and tomatoes which showed a high variability in the content of natural antioxidants and total antioxidant capacity (TRAP) values. For example, kale was found to have the highest TRAP value, followed by lettuce, onions and tomatoes.
- The antioxidant levels and TRAP values of fresh prepared lettuce and onions decreased significantly during chilled (5 °C) storage under all air and MAP conditions. However, the TRAP values of fresh prepared lettuce after one week's chilled storage under high O_2 (80%) MAP were higher compared with fresh prepared lettuce chilled stored under air or low O_2 (5%) MAP.

- In comparison with air packing and low O_2 MAP, high O_2 MAP was not found to preferentially decrease the antioxidant (i.e. ascorbic acid, β-carotene and lutein) levels in fresh prepared lettuce and onions. High O_2 MAP was found to induce the loss of certain phenolic compounds (i.e. quercetin, ferulic acid and cumaric acid), but not luteolin, in fresh prepared lettuce and onions, even though desirable TRAP values were higher after chilled storage compared with air packing and low O_2 MAP.

- The antioxidant efficiency of protection against lipid peroxidation was marginally lower in fresh prepared lettuce and onion extracts stored under high O_2 MAP in comparison with air packing or low O_2 MAP.

- Extracts from high O_2 MA packed fresh prepared lettuce and onions did not have any cytotoxic effects on human colon cells.

- Ingestion of fresh lettuce resulted in increases in human plasma TRAP values through the absorption of phenolic compounds and single antioxidant molecules. These increases in human plasma TRAP values were significantly higher after ingestion of fresh lettuce compared with ingestion of lettuce that had been chilled (5°C) stored for three days.

- Ingestion of chilled stored lettuce packed under air and high O_2 MAs resulted in measurable increases in human plasma TRAP values, whereas virtually no increases in TRAP values were measured after ingestion of equivalent lettuce packed under low O_2 MAs (Figure 8).

Figure 8: Human plasma total antioxidant capacity (TRAP) values after ingestion of fresh lettuce and chilled stored (5 °C for 3 days) air, low O_2 and high O_2 MA packed lettuce

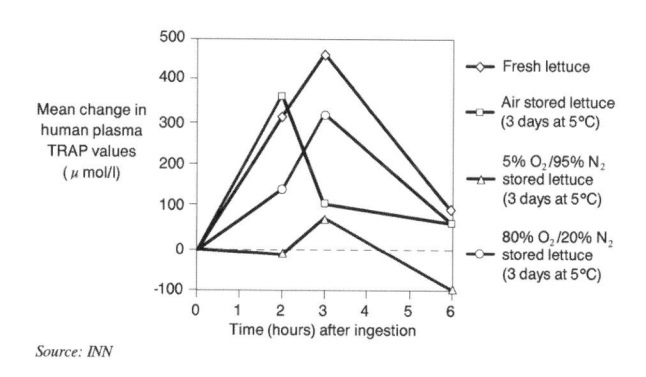

Source: INN

5. FUTURE RESEARCH DIRECTIONS

Specifically with regard to the high O_2 MAP of fresh prepared produce, the following future research directions are suggested:

- Further investigate the potential applications of an innovative dual-action O_2 emitter/CO_2 scavenger active packaging sachet that has been recently developed by Standa Industrie (Caen, France) and marketed by Emco Packaging Systems (Worth, Kent, UK). Initial trials carried out by CCFRA and LinPac Plastics Limited (Pontefract, Yorkshire, UK) in association with several soft fruit suppliers have clearly demonstrated the shelf-life extending potential of this active packaging device (McGrath, 2000). This O_2 emitter/CO_2 scavenger sachet enables high O_2 levels to be maintained within high O_2 MA packs of respiring fresh prepared produce whilst simultaneously controlling CO_2 below levels that may cause physiological damage to produce. Also, the inclusion of this sachet within high O_2 MA packs of fresh prepared produce that have a high intrinsic respiration rate and/or produce volume/gas volume ratio will prevent excessive depletion of in-pack O_2 levels and build-up of in-pack CO_2 levels. In addition, this sachet could also be utilised in low O_2 MA packs of fresh prepared produce to prevent the development of undesirable anaerobic conditions during chilled storage.
- Thoroughly investigate the potential synergy between high O_2 MAP and other active packaging devices (e.g. moisture absorbers, ethylene scavengers and antimicrobial films) and suitable edible coatings and films (Day, 1994; Baldwin *et al*, 1995; Nussinovitch and Lurie, 1995; Rooney, 1999; CCFRA, 2000). Selection criteria of promising active packaging devices and edible coatings and films should be based on their technical efficacy, cost, regulatory status and consumer acceptability (CCFRA, 2000; Day, 2000a).
- Carry out further underpinning research investigations on the effects of high O_2 MAP on the various spoilage and pathogenic microorganisms associated with fresh prepared produce items. Also, further research is merited on the effects of high O_2 MAP on the beneficial nutritional components present in fresh produce and on the complex biochemical reactions and physiological processes that occur during storage.

- Establish optimal high O_2 MAP applications for extending the quality shelf-life and assuring the microbial safety of further fresh prepared produce items (see section 3.3.6) and combination food products which consist of respiring fresh prepared produce and non-respiring food items (e.g. ready meals, pizzas, kebabs and sandwiches). Initial trials carried out by CCFRA have already clearly demonstrated that high O_2 MAP is capable of extending the achievable shelf-life of several chilled ready meals, in comparison with CO_2/N_2 MAP and industry-standard air packing (Day *et al*, 1999; Day, 2000b).

With regard to more general aspects of fresh prepared produce, the following knowledge gaps and suggested research directions are highlighted, in order to assist researchers in the future:

- Provision of packaging film gas permeability data on commercial laminations and coextrusions at realistic chilled temperatures (0-10 °C) and relative humidities (85-95%). At the present time, virtually all gas permeability data is quoted for single films at unrealistic storage temperatures and relative humidities (e.g. 23 °C and 0% RH, see Appendix II).
- Provision of extensive respiration rate data on a wide variety of fresh prepared produce items at different chilled temperatures and under various gaseous storage conditions. At the present time, most respiration rate data available is for whole produce items stored in air (see Appendix I).
- Provision of data on the physiological tolerance of fresh prepared produce items to low (and possibly high) O_2 levels and elevated CO_2 levels. Currently, extensive data is available on the tolerance of whole produce items to low O_2 and high CO_2 levels (Kader *et al*, 1989) but there is a dearth of information on the tolerance of fresh prepared produce items to varying gaseous levels.
- Provision of information on the residual effects of MAP on individual fresh prepared produce items after subsequent pack opening and storage in air.
- Thoroughly investigate an integrated approach to minimal processing techniques, which cover the entire chain from "farm to fork", so as to maintain the sensory quality and assure the microbial safety of fresh prepared produce (Ahvenainen, 1996).
- Carry out further investigations on new and innovative natural preservatives, such as those produced by lactic acid bacteria and those derived from herbs and spices (Kets, 1999).

- Devise improved washing and decontamination procedures for fresh prepared produce that are based on safe non-chlorine alternatives.
- Develop peeling and cutting machinery that can process fresh produce more gently and hence extend the quality shelf-life of fresh prepared produce.
- Devote more resources into refrigeration equipment, design and logistics so that optimal storage temperatures (i.e. 0-3°C) for fresh prepared produce can be maintained throughout the entire chill chain.

REFERENCES

Ahvenainen, R. (1996) New approaches in improving the shelf-life of minimally processed fruit and vegetables. Trends in Food Science and Technology 7 (6), 179-187.

Air Products (1995) The Freshline(r) guide to modified atmosphere packaging (MAP). Air Products Plc, Basingstoke, Hants., UK, pp 1-66.

Anon. (1991) Sulphites banned. Food Ingredients and Processing International 11, 11.

Anon. (2000) What MAP can and cannot do for you. Fresh-cut 8 (6), 16-28.

Baldwin, E.A., Nisperos-Carriedo, M.O. and Baker, R.A. (1995) Use of edible coatings to preserve quality of lightly (and slightly) processed products. Critical Reviews in Food Science and Nutrition 35, 509-524.

Barry-Ryan, C. and O'Beirne, D. (1999) Ascorbic acid retention in minimally processed iceberg lettuce. Journal of Food Science 64, 498-500.

BCGA (1998) The safe application of oxygen enriched atmospheres when packaging food. British Compressed Gases Association Guidance Note GD5, BCGA, Eastleigh, Hants., UK, pp 1-39.

Betts, G.D. (1996) A code of practice for the manufacture of vacuum and modified atmosphere packaged chilled foods. Guideline No. 11, Campden & Chorleywood Food Research Association, Chipping Campden, Glos., UK.

Block, G., Patterson, B. and Subar, A. (1992) Fruit, vegetables and cancer prevention: a review of the epidemiological evidence. Nutrition and Cancer 18, 1-29.

Brecht, J.K. (1995) Physiology of lightly processed fruits and vegetables. HortScience 30 (1), 18-22.

Brody, A.L. and Thaler, M.C. (1996) Argon and other noble gases to enhance modified atmosphere food processing and packaging. Proceedings of IoPP conference on "Advanced Technology of Packaging", Chicago, Illinois, USA, 17th November.

Burns, J.K. (1995) Lightly processed fruits and vegetables: Introduction to the colloquium. HortScience 30 (1), 14.

CCFRA (1999a) Enlabel - a multimedia guide to food labelling legislation. CD-ROM. Campden & Chorleywood Food Research Association, Chipping Campden, Glos., UK.

CCFRA (1999b) Focus on prepared vegetable, salad and fruit products in the UK. Marketplace Report, Campden & Chorleywood Food Research Association, Chipping Campden, Glos., UK, pp 1-37.

CCFRA (2000) Proceedings of the International Conference on "Active and Intelligent Packaging" (ed. B.P.F. Day), Campden & Chorleywood Food Research Association, Chipping Campden, Glos., UK.

CFA (1997) Guidelines for good hygienic practice in the manufacture of chilled foods. 3rd edition, Chilled Foods Association, London, UK.

Chambers, B.J. (1999) Good agricultural practice for fresh produce. In: Proceedings of the International Conference on "Fresh-cut Produce", Campden & Chorleywood Food Research Association, Chipping Campden, Glos., UK.

CHGL (1999) Food Safety (General Food Hygiene) Regulations 1995 and Food Safety (Temperature Control) Regulations 1995: Industry guide to good hygiene practice: fresh produce. Chadwick House Group Limited, London, UK.

Cooper, C. (1999) Issues impacting the fresh-cut produce industry - a US perspective. In: Proceedings of the International Conference on "Fresh-cut Produce", Campden & Chorleywood Food Research Association, Chipping Campden, Glos., UK.

Day, B.P.F. (1992) Guidelines for the good manufacturing and handling of modified atmosphere packed food products. Technical Manual No. 34, Campden & Chorleywood Food Research Association, Chipping Campden, Glos., UK.

Day, B.P.F. (1994) Modified atmosphere packaging and active packaging of fruits and vegetables. In: Minimal Processing of Foods, VTT Symposium Series 142, VTT, Espoo, Finland, pp 173-207.

Day, B.P.F. (1998) Novel MAP - a brand new approach. Food Manufacture 73 (11), 22-24.

Day, B.P.F. (2000a) Consumer acceptability of active and intelligent packaging. Proceedings of the Conference on "Active and Intelligent Packaging: ideas for tomorrow or solutions for today", TNO Nutrition and Food Research, Zeist, The Netherlands.

Day, B.P.F. (2000b) Novel MAP for freshly prepared fruit and vegetable products. Postharvest News and Information 11 (3), 27N-31N.

Day, B.P.F., Bankier, W.J. and Evans, J.M. (1999) Novel high oxygen modified atmosphere packaging (MAP) for chilled combination food products. Research Summary Sheet No. 49, Campden & Chorleywood Food Research Association, Chipping Campden, Glos., UK.

Day, B.P.F. and Gorris, L.G.M. (1993) Modified atmosphere packaging of fresh produce in the Western European market. ZFL - International Food Manufacturing 44, 32-37.

Day, B.P.F. and Wiktorowicz, R. (1999) MAP goes on-line. Food Manufacture 74 (6), 40-41.

DeRoever, C. (1998) Microbiological safety evaluations and recommendations on fresh produce. Food Control 9, 321-347.

Duncan, E. (1999) Non-sulphite dips for fresh prepared produce. In: Proceedings of the International Conference on "Fresh-cut Produce", Campden & Chorleywood Food Research Association, Chipping Campden, Glos., UK.

Elphick, A. (1998) Fruit and vegetable washing systems. Food Processing 67 (1), 22-23.

Exama, A., Arul, J., Lencki, R.W., Lee, L.Z., and Toupin, C. (1993) Suitability of plastic packaging films for modified atmosphere packaging of fruits and vegetables. Journal of Food Science 58 (6), 1365-1370.

Fath, D. and Soudain, P. (1992) Method for the preservation of fresh vegetables. US Patent No. 5128160.

FDA (1998) Guide to minimize microbial food safety hazards for fresh fruits and vegetables. Food and Drug Administration, Washington D.C., USA, pp 1-40.

Ferro-Luzzi, A., Maiani, G. and Catasta, G. (1999) Flavonoids of the Mediterranean diet. Atherosclerosis 144 (1), 169-170.

Francis, G.A., Thomas, C. and O'Beirne, D. (1999) The microbiological safety of minimally processed vegetables. International Journal of Food Science and Technology 34, 1-22.

Garrett, E.H. (1994) Challenges and opportunities in marketing fresh-cut produce. In: Modified Atmosphere Food Packaging (ed. A.L. Brody). Institute of Packaging Professionals, Herndon, Virginia, USA, pp 31-34.

Garrett, E.H. (1998) Fresh-cut produce. In: "Principles and applications of modified atmosphere packaging of foods", 2nd edition (ed. B.A. Blakistone), Blackie Academic & Professional, London, UK, pp 125-134.

Geeson, J. (1999) The application of microperforated films for MAP of fresh prepared produce. In: Proceedings of the International Conference on "Fresh-cut produce", Campden & Chorleywood Food Research Association, Chipping Campden, Glos., UK.

Gonzalez Roncero, M.I. and Day, B.P.F. (1998) The effects of novel MAP on fresh prepared produce microbial growth. Proceedings of the Cost 915 Conference, Ciudad Universitaria, Madrid, Spain, 15-16th October.

Hartley, D.R. (2000) The product design perspective on fresh produce packaging. Postharvest News and Information 11 (3), 35N-38N.

Heard, G. (2000) Microbial safety of ready-to-eat salads and minimally processed vegetables and fruits. Food Science and Technology Today 14 (1), 15-21.

Holdsworth, S.D. (1983) The preservation of fruit and vegetable food products. Science in Horticulture Series (ed. L. Broadbent), MacMillan Press, London, UK, pp 61-98.

Hurst, W.C. (1995) Sanitation of lightly processed fruits and vegetables. HortScience 30 (1), 22-24.

IFPA (1993) Assessment of the risk of botulism contributed by modified atmosphere packaging of fresh-cut produce. International Fresh-cut Produce Association, Alexandria, Virginia, USA.

IFPA (1996) Food safety guidelines for the fresh-cut produce industry. 3rd edition, International Fresh-cut Produce Association, Alexandria, Virginia, USA.

IFPA (1999) Fresh-cut produce guidelines. 3rd edition, International Fresh-cut Produce Association, Alexandria, Virginia, USA.

IFPA (2000) HACCP for the fresh-cut produce industry. 4th edition, International Fresh-cut Produce Association, Alexandria, Virginia, USA, pp 1-13.

Kader, A.A. and Ben-Yehoshua, S. (2000) Effects of superatmospheric oxygen levels on postharvest physiology and quality of fresh fruits and vegetables. Postharvest Biology and Technology 20 (2000), 1-13.

Kader, A.A., Zagory, D. and Kerbel, E.L. (1989) Modified atmosphere packaging of fruits and vegetables. Critical Reviews in Food Science and Nutrition 28 (1), 1-30.

Kets, E.P.W. (1999) Applications of natural anti-microbial compounds. In: Proceedings of the International Conference on "Fresh-cut Produce", Campden & Chorleywood Food Research Association, Chipping Campden, Glos., UK.

Keynote (2000) Fruit and vegetables market report. Keynote Ltd., Hampton, Middlesex, UK.

Knight, C. (1999) HACCP for fresh produce. In: Proceedings of the International Conference on "Fresh-cut Produce", Campden & Chorleywood Food Research Association, Chipping Campden, Glos., UK.

Laurila, E., Kervinen, R. and Ahvenainen, R. (1998) The inhibition of enzymatic browning in minimally processed vegetables and fruits. Postharvest News and Information 9 (4), 53-66.

Leaper, S. (1997) HACCP: a practical guide. Technical Manual No. 38, 2nd edition, Campden & Chorleywood Food Research Association, Chipping Campden, Glos., UK.

MAFF (1996) EC quality standards for horticultural produce (fresh vegetables, fruits, salads and flowers). Ministry of Agriculture, Fisheries and Food, London, UK.

Marston, E.V. (1995) Suitability of films for MAP of fresh produce. Produce Technology Monitor 5 (7), 1-2.

Martens, T. (1999) Harmonization of safety criteria for minimally processed foods. Final report of the FAIR CT96-1020 concerted action project. European Commission, Brussels, Belgium.

McEvily, A.J., Iyengar. R. and Otwell, W.S. (1992) Inhibition of enzymatic browning in foods and beverages. Critical Reviews in Food Science and Nutrition, 32 (3), 253-273.

McGrath, P. (2000) Smart fruit packaging. Grower 133 (22), 15-16.

Miller, A.R. (1992) Physiology, biochemistry and detection of bruising (mechanical stress) in fruits and vegetables. Postharvest News & Information 3, 53-58.

Ness, A.R. and Powles, J.W. (1997) Fruit and vegetables, and cardiovascular disease: a review. International Journal of Epidemiology 26, 1-13.

Nguyen-the, C. and Carlin, F. (1994) The microbiology of minimally processed fresh fruits and vegetables. Critical Reviews in Food Science and Nutrition 34, 371-401.

Nussinovitch, A. and Lurie, S. (1995) Edible coatings for fruits and vegetables. Postharvest News and Information 6(4), 53N-57N.

O'Beirne, D. (1999) Modified atmosphere packed vegetables and fruit - an overview. In: Proceedings of the International Conference on "Fresh-cut Produce", Campden & Chorleywood Food Research Association, Chipping Campden, Glos., UK.

Puupponen-Pimiä, R. (1999) Bioactive compounds in fruit and vegetables - a review. In: Proceedings of the International Conference on "Fresh-cut Produce", Campden & Chorleywood Food Research Association, Chipping Campden, Glos., UK.

Romig, W.R. (1995) Selection of cultivars for lightly processed fruits and vegetables. HortScience 30 (1) 38-40.

Rooney, M. (1999) Active and intelligent packaging of fruit and vegetables. In: Proceedings of the International Conference on "Fresh-cut Produce", Campden & Chorleywood Food Research Association, Chipping Campden, Glos., UK.

Rose, D. (1994) Risk factors associated with post process contamination of heat sealed semi-rigid packaging. Part 2A: Investigation of microbial ingress through laser drilled microholes in packaging. Technical Memorandum 708, Campden & Chorleywood Food Research Association, Chipping Campden, Glos., UK.

Sapers, G.M. (1993) Browning of foods: control by sulfites, oxidants and other means. Food Technology 47 (10), 75-84.

Schlimme, D.V. (1995) Marketing lightly processed fruits and vegetables. HortScience 30 (1), 15-17.

Serafini, M. (1999) Nutritional and health promoting benefits of fruit and vegetable consumption. In: Proceedings of the International Conference on "Fresh-cut Produce", Campden & Chorleywood Food Research Association, Chipping Campden, Glos., UK.

Seymour, I.J. (1999) Review of current industry practice on fruit and vegetable decontamination. Review No. 14, Campden & Chorleywood Food Research Association, Chipping Campden, Glos., UK.

Simons, L.K. and Sanguansri, P. (1997) Advances in the washing of minimally processed vegetables. Food Australia 49, 75-80.

Spencer, K. (1999) Fresh-cut produce - applications of noble gases. In: Proceedings of the International Conference on "Fresh-cut Produce", Campden & Chorleywood Food Research Association, Chipping Campden, Glos., UK.

Thom, S.R. and Marquis, R.E. (1984) Microbial growth modification by compressed gases and hydrostatic pressure. Applied and Environmental Microbiology 47 (4), 780.

Whitaker, J.R. (1996) Enzymes. In: "Food Chemistry", 3rd edition (ed. O.R. Fennema), Marcel Dekker, Inc., New York, USA, pp 493-496.

WHO (1998) Surface decontamination of fruits and vegetables eaten raw: a review. World Health Organisation (Food Safety Unit), Washington D.C., USA, pp 1-42.

Wiley, R.C. (1994) Minimally processed refrigerated fruits and vegetables. Chapman & Hall, New York, USA.

APPENDIX I

Reported respiration rates of whole and prepared fresh produce items in air

Fresh Produce Items [a]	Respiration rate range (ml CO_2/kg/h) [b]							
	0 °C		5 °C		10 °C [c]		20-24 °C	
	Min	Max	Min	Max	Min	Max	Min	Max
Apple	1 [8]	3 [14]	2.6 [14]	5.7 [14]	3.6 [8]	10.7 [14]	7.4 [8]	31.6 [10]
Apricot	2.5 [14]	3 [14]	3.1 [14]	4.7 [14]	5.9 [14]	10.2 [14]	16.4 [14]	29.4 [14]
Artichokes, globe	7.6 [14]	22.7 [14]	13.5 [14]	31.1 [14]	29 [14]	52.4 [14]	76.3 [14]	132 [14]
Asparagus	13.6 [14]	40.4 [14]	22.8 [1]	70.5 [14]	33.7 [1]	163 [14]	71.8 [1]	283 [14]
Aubergine	3.8 [3]		6 [3]		10.5 [3]		45.2 [3]	
Avocado			10.4 [14]	15.5 [14]			41.8 [14]	196 [14]
Banana without peel:								
Whole	3.5 [11]		4.7 [11]		5.8 [11]		52.9 [11]	
Sliced	4 [11]		5.4 [11]		11.3 [11]		67.2 [11]	
Beans:								
Broad	17.7 1		26.9 1		46.5 1		81.9 1	
Green:								
Whole	6.6 [11]		15 [11]	17.5 [8]	27.8 [11]	28.3 [8]	59.6 [8]	74 [11]
Cut	7 [11]		15 [11]		41.7 [11]		88.1 [11]	
Runner	10.6 [1]		14.5 [1]		19.3 [1]		50.8 [1]	
Snap	10.1 [14]		18.1 [14]		31 [14]		73.4 [14]	
Bean sprouts	10.6 [14]	12.6 [14]	21.8 [14]		49.7 [14]	53 [14]		
Blackberry	9.1 [14]	11 [1]	10 [8]	21.2 [14]	33.2 [1]		87.6 [1]	
Blackcurrant	8.1 [1]		14 [1]		20.9 [1]		73.4 [1]	
Blueberry	1 [14]	5.1 [14]	4.7 [14]	6.2 [14]	12.3 [14]	18.7 [14]	29.4 [14]	49.2 [14]
Broccoli:								
Whole	10 [8]	38.9 [1]	21 [8]	30 [8]	59 [5]	85 [8]	104 [5]	240.1 [1]
Cut florets					78 [5]		147 [5]	
Brussel sprouts:								
Trimmed					25 [5]		56 [5]	
Untrimmed	8.6 [1]		15.5 [1]	30 [8]	25 [5]	26.7 [1]	50.8 [1]	72 [5]

Fresh Produce Items [a]	Respiration rate range (ml CO_2/kg/h) [b]							
	0 °C		5 °C		10 °C [c]		20-24 °C	
	Min	Max	Min	Max	Min	Max	Min	Max
Cabbage:								
Primo	5.6 [1]		13.5 [1]		16 [1]		22.6 [1]	
January King	3 [1]		6.7 [1]		13.9 [1]		32.2 [1]	
Decema	1.5 [1]		3.6 [1]		4.3 [1]		11.3 [1]	
Chinese:								
Half	5 [2]		8 [2]		9 [2]		17 [2]	20 [2]
Rough shred (0.5 x 3 cm)	9 [2]		16 [2]		25 [2]		50 [2]	55 [2]
Fine shred (0.25 x 1.5 cm)	12 [2]		20 [2]		30 [2]		65 [2]	70 [2]
Apex:								
Quarters	4 [2]		6 [2]		12 [2]		35 [2]	
Rough shred (1 x 3 cm)	9 [2]		14 [2]	17 [2]	25 [2]		80 [2]	85 [2]
Fine shred (0.5 x 1.5 cm)	12 [2]		20 [2]		30 [2]		95 [2]	
Lenox:								
Quarters	4 [2]		5 [2]		10 [2]		30 [2]	
Rough shred (1 x 3 cm)	8 [2]		11 [2]		22 [2]		65 [2]	70 [2]
Fine shred (0.5 x 1.5 cm)	9 [2]		13 [2]		27 [2]		85 [2]	
Carrots:								
Unpeeled/storing	2.5 [3]	6.6 [1]	5 [8]	10 [8]	9 [5]	20 [9]	18.6 [1]	29 [5]
Bunching with leaves	9.1 [14]	17.7 [1]	12.9 [14]	26.4 [1]	17.1 [14]	39.6 [1]	26 [14]	68.4 [1]
Peeled, whole			5 [12]	6 [12]	9 [12]	11 [12]	26 [5]	30 [12]
Julienne cut	7.6 [12]		9.7 [12]		24.3 [12]	65 [5]	145 [5]	
Grated	5.7 [12]		12.1 [12]	13 [2]	22.1 [12]	27 [2]	60 [2]	70 [2]
Cauliflower	8.1 [14]	14.7 [3]	9.8 [14]	29.5 [3]	17.1 [14]	45.6 [3]	42.4 [14]	108.5 [3]
Celeriac	3.5 [14]		7.8 [14]		13.4 [14]		28.2 [14]	
Celery	3.5 [1]		4.7 [1]	5.7 [14]	6.4 [1]	12.8 [14]	18.6 [1]	36 [10]

Fresh Produce Items [a]	Respiration rate range (ml CO_2/kg/h) [b]							
	0 °C		5 °C		10 °C [c]		20-24 °C	
	Min	Max	Min	Max	Min	Max	Min	Max
Cherry:								
Sweet	2 [14]	2.5 [14]	5.2 [14]	7.3 [14]			15.8 [14]	18.1 [14]
Sour	3 [14]	6.6 [14]	6.7 [14]				22 [14]	28.2 [14]
Chilli Pepper	34 [8]		4 [8]		5 [8]			
Courgette:								
Whole	6.6 [11]		15.5 [11]		30.5 [11]		81.4 [11]	
Slices	6.1 [11]		12.4 [11]		25.1 [11]		91 [11]	
Cranberry	2.1 [14]	2.6 [14]					6.2 [14]	10.2 [14]
Cucumber:								
Whole	1.4 [11]	3 [1]	2.2 [11]	4.1 [1]	3.5 [11]	7 [1]	8.5 [1]	
Slices	1.7 [11]		2.8 [11]		5.2 [11]		25.4 [11]	
Dill	22.2 [3]		32 [3]		51.6 [3]		131.9 [3]	
Endive	22.7 [14]		26.9 [14]		53.5 [14]		75.1 [14]	
Figs			5.7 [14]	6.7 [14]	11.8 [14]	12.3 [14]	32.2 [14]	53.7 [14]
Garlic	5.3 [3]		13.3 [3]		26.4 [3]		79.3 [3]	
Gherkin					10 [9]	20 [9]		
Gooseberry	2.5 [14]	5.1 [1]	4.1 [14]	8.3 [14]	6.4 [14]	17.1 [14]	23.2 [14]	59.3 [14]
Grape (green seedless):								
With stem	1.1 [11]		1.7 [11]		2.9 [11]		11 [11]	
Without stem	1 [11]		1.5 [11]		3 [11]		12.6 [11]	
Kiwi Fruit:								
Whole	1.6 [11]		2.4 [11]		4.6 [11]		12.4 [11]	
Sliced	3.6 [11]		6 [11]		12.5 [11]		43.9 [11]	
Leek:								
Whole leaf	10.1 [1]	12 [2]	14.5 [1]	15 [2]	20 [2]	26.7 [1]	62.1 [1]	65 [2]
_ leaf + _ stem	16 [2]		20 [2]		30 [2]	35 [2]	80 [2]	90 [2]
Rings (2mm)	22 [2]		25 [2]		70 [2]		140 [2]	160 [2]
Lemon					5.9 [14]		10.7 [14]	14.7 [4]
Lettuce:								
Kale	8.1 [14]	13.6 [14]	17.6 [14]	24.4 [14]	38.5 [14]	44.9 [14]	105 [14]	150 [14]
Kohirabi	5.1 [14]		8.3 [14]		16.6 [14]			

High oxygen MAP for fresh prepared produce

Fresh Produce Items [a]	Respiration rate range (ml CO_2/kg/h) [b]							
	0 °C		5 °C		10 °C [c]		20-24 °C	
	Min	Max	Min	Max	Min	Max	Min	Max
Kordaat	4.5 [1]		5.7 [1]		9.1 [1]		20.9 [1]	
Kloek	8.1 [1]		12.4 [1]		16.6 [1]		45.2 [1]	
Unrivalled	9.1 [1]		5.9 [1]		7.4 [1]		48 [1]	
Iceberg:								
Quarters					9 [6]			
Shred (washed)	6 [12]	7.5 [12]	8 [12]	14 [12]	16 [12]	21 [12]		
Shred (unwashed)					22 [6]			
Lima Bean	5.1 [14]	15.2 [14]	10 [8]	20 [8]			75.1 [14]	101 [14]
Lime (Tahiti)							4 [14]	10.8 [14]
Mango			5.2 [14]	11.4 [14]	15		42.4 [14]	85.3 [14]
Melon (Cantaloupe)	2.5 [14]	3 [14]	4.7 [14]	5.2 [14]	7.5 [14]	8.6 [14]	25.4 [14]	36.7 [14]
Melon (Honeydew):								
Whole	0.7 [11]		1.6 [14]	2.6 [14]	2.8 [11]	4.8 [14]	5.6 [11]	15.3 [14]
Cubes	1.2 [11]		1.55 [11]		4.4 [11]		35 [11]	
Melon (Watermelon)			1.6 [14]	2.1 [14]	3.2 [14]	4.8 [14]	9.6 [14]	14.1 [14]
Mushroom:								
Sliced (5mm)	10 [12]	30 [12]	20 [12]	45 [12]	45 [12]	65 [12]	191 [6]	
Whole	12.1 [3]		26.4 [3]		48.1 [3]	53 [6]	126.6 [3]	153 [6]
Okra			27.4 [14]	30.6 [14]	46 [14]	51 [14]	140 [14]	155 [14]
Olives							22.6 [14]	59.3 [14]
Onion:								
Whole	1.5 [1]	4 [2]	2.6 [1]	5 [8]	3.7 [1]	11 [2]	4.5 [1]	33 [2]
Rings (2 mm)	7 [2]		12 [2]		20 [2]		70 [2]	73 [2]
Diced (1cm x 1cm)	6 [2]		8 [2]		12 [2]		50 [2]	55 [2]
Green	5.1 [14]	16.2 [14]	8.8 [14]	30 [8]	19.3 [14]	40.1 [3]	37.3 [14]	79.1 [3]
Orange	1 [14]	2.5 [14]	2.1 [14]	3.6 [14]	3.2 [14]	4.8 [14]	12.4 [14]	19.2 [14]
Papaya			2.1 [14]	3.1 [14]				
Parsley	15.2 [14]	31.8 [3]	27.5 [14]	47.2 [3]	45.5 [14]	87.7 [14]	111 [14]	165.5 [3]
Parsnip	3.5 [1]	7.6 [14]	4.7 [14]	9.3 [14]	10.7 [14]	13.9 [14]	27.7 [1]	
Peach (sliced)	2 [14]	6 [12]	3.1 [14]	23 [12]	8.6 [14]	53 [12]	33.3 [14]	57.7 [14]

© CCFRA 2001

High oxygen MAP for fresh prepared produce

Fresh Produce Items [a]	Respiration rate range (ml CO_2/kg/h) [b]							
	0 °C		5 °C		10 °C [c]		20-24 °C	
	Min	Max	Min	Max	Min	Max	Min	Max
Pear:								
Bartlett	1.5 [14]	3.5 [14]	2.6 [14]	5.2 [14]	4.3 [14]	11.2 [14]	17 [14]	39.5 [14]
Kieffer	1 [14]						8.5 [14]	15.8 [14]
Peas (in pod):								
Early	20.2 [1]		31.6 [1]		69.5 [1]		144.1 [1]	
Maincrop	23.7 [1]		16.2 [3]	28.5 [1]	32.5 [3]	64.2 [1]	91.1 [3]	141.2 [1]
Pepper (Capsicum):								
Whole	2.4 [3]	4.1 [1]	4.1 [11]	10.4 [8]	7 [11]	21.4 [9]	19.8 [1]	44.6 [7]
Sliced	3.5 [11]		3.1 [11]		7.5 [11]		59.3 [11]	
Pineapple:								
Mature Green (Chunks)			1 [14]	2.5 [12]	2.1 [14]	7 [12]	10.7 [14]	16.4 [14]
Golden (Chunks)			5.5 [12]	7 [12]	13	16 [12]		
Plum (Wickson)	1 [14]	1.5 [14]	2.1 [14]	4.7 [14]	3.7 [14]	5.9 [14]	10.2 [14]	14.7 [14]
Potato:								
Bintje:								
Whole, peeled	3 [2]		4 [2]		9 [2]		35 [2]	
Half			4 [2]		12 [2]		45 [2]	
Sliced (2mm)	5 [2]		6 [2]		20 [2]		70 [2]	
Van Gogh:								
Whole, peeled	3 [2]		4 [2]		10 [2]		30 [2]	
Half	4 [2]		5 [2]		1 [2]		45 [2]	
Sliced (2mm)	6 [2]		8 [2]		20 [2]		65 [2]	
King Edward:								
Maincrop	3 [1]		1.6 [1]		2.1 [1]		3.4 [1]	
New	5.1 [1]		7.8 [1]		10.7 [1]		22.6 [1]	
Jersey Royal:								
Whole, peeled					12 [13]		90 [13]	
Pentland Squire:								
Whole					9 [13]			
Chipped					18 [13]			

High oxygen MAP for fresh prepared produce

Fresh Produce Items [a]	Respiration rate range (ml CO_2/kg/h) [b]							
	0 °C		5 °C		10 °C [c]		20-24 °C	
	Min	Max	Min	Max	Min	Max	Min	Max
Radish:								
With tops	7.1 [14]	8.6 [14]	9.8 [14]	12.1 [3]	16.6 [14]	37.4 [9]	46.6 [3]	76.8 [14]
Topped	1.5 [14]	4.5 [14]	3.1 [14]	6.7 [14]	8 [14]	8.6 [14]	25 [14]	32.8 [14]
Raspberry	9.1 [14]	12.6 [14]	10 [8]	28.5 [1]	15 [14]	49.2 [1]	123 [1]	
Red Beet:								
Whole, peeled	2 [2]		3 [2]		10 [2]		30 [2]	
Cubed (1 x 1cm)	5 [2]		6 [2]		14 [2]		60 [2]	70 [2]
Grated (2mm shred)	6 [2]		8 [2]		20 [2]		90 [2]	115 [2]
Unwashed, grated	8 [2]		11 [2]		25 [2]		140 [2]	155 [2]
Rutabega:								
Quarters	4 [2]		6 [2]		8 [2]		25 [2]	
Cubed (1 x 1cm)	8 [2]		10 [2]		22 [2]		40 [2]	45 [2]
Grated (2mm shred)	9 [2]		13 [2]		22 [2]		85 [2]	90 [2]
Rhubarb	7.1 [1]		10.9 [1]		10 [9]	20 [9]	30.5 [1]	
Spinach	9.6 [14]	25.3 [1]	18.1 [14]	30.1 [14]	35 [9]	73.8 [14]	71.8 [1]	162 [14]
Strawberry	6.1 [14]	9.1 [14]	8.3 [14]	20 [8]	26.2 [14]	50.8 [14]	57.6 [14]	111 [14]
Sweetcorn	15.2 [10]	15.6 [1]	22.3 [10]		48.1 [1]		118.7 [1]	129 [10]
Tomato:								
Whole	0.8 [11]	3 [1]	1.2 [11]	10 [8]	2.5 [11]	10 [9]	11.4 [11]	16.9 [1]
Sliced	0.7 [11]		1.6 [11]		5.3 [11]		19.8 [11]	
Mature green			2.6 [14]	4.1 [14]	6.4 [14]	9.6 [14]	15.8 [14]	23.2 [14]
Ripening					7 [14]	8.6 [14]	13.6 [14]	25 [14]
Turnip	7.6 [1]		8.8 [1]		5 [9]	16 [1]	29.4 [1]	
Watercress	7.6 [14]	13.1 [14]	18.6 [1]	25.4 [14]	42.8 [1]	64.7 [14]	116.9 [1]	197 [14]

[a]   Unless stated, produce items are whole and unprepared. Where known: produce type, varieties and degree of preparation are included.

[b]   mg converted to ml CO_2 using densities of CO_2 at:
0 °C = 1.98;   5 °C = 1.93;   10 °C = 1.87;   20 °C = 1.77

[c]   Relative produce respiration rates (ml CO_2/kg/hr) at 10 °C in air have been classified 15 as follows:
Low   <10 ml CO_2/kg/hr   Very High   40-60 ml CO_2/kg/hr
Medium   10-20 ml CO_2/kg/hr   Extremely High   >60 ml CO_2/kg/hr
High   20-40 ml CO_2/kg/hr

References for Appendix I

1. Robinson, J.E., Browne, K.M and Burton, W.G. (1975) Storage characteristics of some vegetables and soft fruits. Annals of Applied Biology 81 (3), 399-408.

2. Mattila, M., Ahvenainen, R., Hurme, E. and Hyvonen, L. (1995) Respiration rates of some minimally processed vegetables. In: Proceedings of the Cost 94 workshop on "The postharvest treatment of fruit and vegetables: systems and operations for postharvest quality" (ed. J. de Baerdemaeker *et al*), European Commission, Brussels, Belgium, pp 135-145.

3. Burzo, I. (1981) Influence of temperature on respiratory intensity in main vegetable varieties. Postharvest handling of vegetables. Acta Horticulturae 116, 61-64.

4. Biale, J.B. (1966) Respiration of fruits. Handbuch der Pflanzenphysiologie (Reihland) 12 (2), 536.

5. Stark, R. and McLachlan, A. (1985) Modified atmosphere packaging of chilled vegetables. Technical Memorandum No. 41, Campden & Chorleywood Food Research Association, Chipping Campden, Glos., UK.

6. Ballantyne, A., Stark, R. and Selman, J.D. (1988) Modified atmosphere packaging of shredded lettuce. International Journal of Food Science and Technology 23 (3), 267.

7. Platenius, H. (1942) Effect of temperature on the respiration rate and respiratory quotient of some vegetables. Plant Physiology 17, 179.

8. Kader, A.A., Zagory, D. and Kerbel, E.L. (1989) Modified atmosphere packaging of fruits and vegetables. Critical Reviews in Food Science and Nutrition 28 (1), 1-30.

9. Kader, A.A. (1985) An overview of the physiology and biochemical basis of CA effects on fresh horticultural crops. Proceedings of the 4th National CA Research Conference, Raleigh, North Carolina, USA, 23-26 July.

10. Hardenburg, R.E. (1971) Effect of in-package environment on keeping quality of fruits and vegetables. HortScience 6 (3), 198.

11. Watada, A.E., Ko, N.P. and Minott, D.A. (1996) Factors affecting quality of fresh-cut horticultural products. Postharvest Biology and Technology 9 (2), 115-125.

12. Gorny, J.R. (1997) A summary of CA requirements and recommendations for fresh-cut (minimally processed) fruits and vegetables. CA '97 Proceedings Vol 5.

13. Ballantyne, A. (1986) Modified atmosphere packaging of selected prepared vegetables. Technical Memorandum No. 436. Campden & Chorleywood Food Research Association, Chipping Campden, Glos., UK.

14. USDA (1971) The commercial storage of fruits, vegetables, and florist and nursery stock. Handbook No. 66, United States Department of Agriculture, Beltsville, Maryland, USA.

15. Day, B.P.F. (1994) Modified atmosphere packaging and active packaging of fruits and vegetables. In: Minimal Processing of Foods, VTT Symposium Series 142, VTT, Espoo, Finland, pp 173-207.

APPENDIX II

Oxygen and water vapour transmission rates of selected packaging materials for fresh produce

Packaging film [a] (25μm)	Oxygen transmission rate ($cm^3 m^{-2} day^{-1} atm.^{-1}$) 23°C:0% RH [b]	Relative oxygen permeability at 23°C:0% RH	Water vapour transmission rate ($g\ m^{-2}\ day^{-1}$) 38°C:90% RH [b]	Relative water vapour transmission rate at 38°C:90% RH
Aluminium (Al)	<0.1 [1]		<0.1 [1]	Barrier, <10
Ethylene-vinyl alcohol (EVOH)	0.2-1.6 [2]	Barrier	24-120 [2]	Variable
Polyvinylidene chloride (PVdC)	0.8-9.2	<50	0.3-3.2	Barrier, <10
Modified nylon (MXD6)	2.4 [2]		25	Semi-barrier, 10-30
Polyester (PET)	50-100		20-30	Semi-barrier, 10-30
Polyamide (nylon) (PA6)	80 [2]	Semi-barrier	200	Very high, 200-300
Modified polyester (PETG)	100	50-200	60	Medium, 30-100
Metallised orientated polypropylene (MOPP)	100-200 [1]		1.5-3.0 [1]	Barrier, <10
Polyvinyl chloride (plasticised) (PVC)	2000-5000 [3]		200 [3]	Very high, 200-300
Orientated polypropylene (OPP)	2000-2500		7	Barrier, <10
High density polyethylene (HDPE)	2100	Medium 200-5000	6-8	Barrier, <10
Polystyrene (PS)	2500-5000		110-160	High, 100-200
Orientated polystyrene (OPS)	2500-5000		170	High, 100-200
Polypropylene (PP)	3000-3700		10-12	Semi-barrier, 10-30
Polycarbonate (PC)	4300		180	Very high, 100-200
Low density polyethylene (LDPE)	7100	High 5000-10000	16-24	Semi-barrier, 10-30
Polyvinyl chloride (highly plasticised) (PVC)	5000-10000 [3]		200 [3]	Very high, 200-300
Ethylene-vinyl acetate (EVA)	12000	Very high 10000-15000	110-160	Very high, 100-200
Microperforated (MP)	>15000 [4]	Extremely high >15000	Variable [4]	Extremely high, >300
Microporous (MPOR)	>15000 [4]		Variable [4]	Extremely high, >300

[a] It should be noted that most packaging films for fresh produce are not single films but laminates and co-extrusions
[b] It should be noted that conditions of O_2 and water vapour transmission rate measurements are not at realistic chill conditions
[1] Dependent on pinholes
[2] Dependent on moisture
[3] Dependent on moisture and level of plasticiser
[4] Dependent on film and degree of microperforation or microporosity

APPENDIX III

Graded sensory quality specification for fresh prepared iceberg lettuce

Appearance

Bright, glossy and crisp in appearance, slightly moist. No discoloration.	Grade A
Some loss of brightness and moisture, slightly limp. May have a very slight amount of discoloration.	Grade B
Slightly dull, moderately dry, moderately limp. May have a slight amount of discoloration.	Grade C
Moderately dull and dry, moderately limp. May have a moderate amount of discoloration. May have slight slimy areas at edges of leaves.	Grade D
Very dull, very limp. May have marked discoloration and/or moderate slimy areas.	Grade E

Odour

Characteristic fresh odour of the variety. No off-odours or excessive bitter notes.	Grade A
Loss of characteristic fresh odour. May have some very slight hay notes.	Grade B
Slight stale hay/compost odour. May have some acid/sour notes.	Grade C
Rotting vegetable odour. May have fermenting and/or strong sour notes.	Grade D

Texture (by mouth)

Crisp texture.	Grade A
Slight loss of crispness. May be very slightly dry.	Grade B
Slightly limp texture. May be slightly dry.	Grade C
Limp texture. May be either dry or have slimy areas.	Grade D

APPENDIX IV

Summary tables of the results from CCFRA's fresh prepared produce trials

Table 1: Summary of fresh prepared iceberg lettuce trials #1-13 (July 1995 – June 1997)

Trial Number	Source	Harvest to Pack Time	Storage Temp.	Gas Analyses (After 10 Days) Ind. Standard[a] $O_2\%$	Ind. Standard[a] $CO_2\%$	High O_2 MAP $O_2\%$	High O_2 MAP $CO_2\%$	Sensory Effects of High O_2 MAP cf. Industry Standard Low O_2 MAP[a]	Micro Effects of High O_2 MAP cf. Industry Standard Low O_2 MAP[a]
1	UK	1 day	3-5°C	<1	5	55[c]	15[c]	very good	very good
2	Spanish	4 days	5°C	3	5	40[c]	14[c]	good/variable	good
3	USA	4 days	6-8°C	<1	9	47[b]	30[b]	variable	variable
4-6	Spanish	3 days	6-8°C	2	10	50[b]	30[b]	good/variable	not determined
7-9	UK	3-4 days	8°C	1	10	40[b]	40[b]	negative/variable	not determined
10-12	UK	1-4 days	8°C	<1	6	35[b]	35[b]	negative/variable	not determined
13	UK	1 day	5°C	2	8	52[d]	13[d]	very good	not determined
			8°C	3	8	47[d]	14[d]	very good	not determined

[a] Industry standard low O_2 MAP: 5% O_2/95% N_2 in OPP film
[b] High O_2 MAP: 80% O_2/20% N_2 in PVDC-coated OPP film
[c] High O_2 MAP: 70% O_2/30% N_2 in PVDC-coated OPP film
[d] High O_2 MAP: 80% O_2/20% N_2 in OPP film

High oxygen MAP for fresh prepared produce

Table 2: Summary of fresh prepared potato trials #1-4 (Aug. 1995 – March 1997)

Trial Number	Harvest to Pack Time	Dip Treatment	Storage Temp.	Gas Analysis (After 10 Days) Ind. Standard [a] $O_2\%$	Ind. Standard [a] $CO_2\%$	High O_2 MAP $O_2\%$	High O_2 MAP $CO_2\%$	Sensory Effects of High O_2 MAP cf. Industry Standard Low O_2 MAP	Micro. Effects of High O_2 MAP cf. Industry Standard Low O_2 MAP
1	4 days	Sulphite	5°C	2	6	70 [c]	25 [c]	variable	very good/good
2	~5-6 months	Sulphite level 1	6-8°C	<1	9	48 [b]	32 [b]	variable	good
		Sulphite level 2		1	8	52 [b]	30 [b]	good/variable	good
3	25 days	Sulphite	8°C	1	10	18 [b]	35 [b]	variable	not determined
		Non-sulphite (EPL) [d]		1	11	16 [b]	54 [b]	very good/good	very good/good
		Non-sulphite (Roche) [d]		<1	12	15 [b]	53 [b]	very good/good	good/variable
4	~4-5 months	Sulphite	5°C	1	10	50 [b]	25 [b]	variable	not determined
			8°C	<1	13	30 [b]	42 [b]	variable	not determined
		Non-sulphite (EPL) [d]	5°C	1	13	45 [b]	35 [b]	good	not determined
			8°C	1	14	26 [b]	43 [b]	good	not determined
		Non-sulphite (Hahn) [d]	5°C	1	12	46 [b]	31 [b]	good/marginal	not determined
			8°C	1	14	26 [b]	50 [b]	marginal	not determined

[a] Industry standard low O_2 MAP: 5% O_2/95% N_2 in OPP film
[b] High O_2 MAP: 80% O_2/20% N_2 in PVDC-coated OPP film
[c] High O_2 MAP: 100% O_2 in PVDC-coated OPP film
[d] Proprietary chemical non-sulphite dip formulations and treatments

© CCFRA 2001

High oxygen MAP for fresh prepared produce

Table 3: Summary of fresh prepared produce trials (Sept. 1995 – June 1997)

Produce Item	Source	Harvest to Pack Time	Storage Temp.	Gas Analysis (After 10 Days) Ind. Standard O$_2$%	Ind. Standard CO$_2$%	High O$_2$ MAP O$_2$%	High O$_2$ MAP CO$_2$%	Sensory Effects of High O$_2$ MAP cf. Industry Standard Low O$_2$ MAP	Micro. Effects of High O$_2$ MAP cf. Industry Standard Low O$_2$ MAP
Cos lettuce #1	UK	1 day	5°C	<1[a]	10[a]	38[f]	24[f]	good/variable	good/variable
Beansprouts	UK	1 day	5°C	13[b]	7[b]	42[f]	22[f]	marginal	variable
Honeydew melon	Brazilian	~1 month	5°C	1[a]	16[a]	45[f]	19[f]	good/marginal	good/variable
Sliced mushrooms	UK	1 day	8°C	18[b]	5[b]	<1[e]	65[e]	marginal	good/variable
Broccoli florets	Spanish	8 days	8°C	15[b]	9[b]	3[e]	71[e]	good/marginal	variable
Apple slices[d]	Belgian	~8 months	5°C	<1[a]	21[a]	51[e]	27[e]	negative/marginal	not determined
			8°C	<1[a]	40[a]	33[e]	45[e]	negative/marginal	not determined
Tomato wedges	UK	1 day	5°C	5[a]	7[a]	55[e]	12[e]	good/variable	not determined
			8°C	4[a]	10[a]	51[e]	19[e]	good/variable	not determined
Strawberries	UK	1 day	5°C	21[c]	<1[c]	29[e]	34[e]	negative/marginal	not determined
			8°C	21[c]	<1[c]	2[e]	67[e]	negative	not determined
Baton carrot #1	UK	7 days	5°C	<1[a]	32[a]	<1[f]	61[f]	variable	very good
Baton carrot #2	UK	4 days	6-8°C	11[b]	13[b]	23[e]	17[e]	good/variable	variable
Baton carrot #3	UK	5 days	8°C	17[b]	7[b]	20[e]	53[e]	marginal	marginal

a   Industry standard low O$_2$ MAP: 5% O$_2$/95% N$_2$ in OPP film
b   Industry standard: in air in microperforated OPP film
c   Industry standard: in air in open PVC punnet
d   Dipped in a proprietary non-sulphite solution from EPL
e   High O$_2$ MAP: 80% O$_2$/20% N$_2$ in PVDC-coated OPP film
f   High O$_2$ MAP: 70% O$_2$/30% N$_2$ in PVDC-coated OPP film

© CCFRA 2001

High oxygen MAP for fresh prepared produce

Table 4: Summary of fresh prepared produce screening trials (Oct – Dec 1996)

Produce Item	Storage Temp.	Gas Analysis (After 10 Days) Ind. Standard O$_2$%	Ind. Standard CO$_2$%	High O$_2$ MAP O$_2$%	High O$_2$ MAP CO$_2$%	Sensory Effects of High O$_2$ MAP cf. Industry Standard Low O$_2$ MAP
Tomato slices	5°C	18 [b]	1 [b]	48 [d]	15 [d]	good
White grapes	5°C	12 [c]	9 [c]	69 [d]	6 [d]	marginal
Diced onions	5°C	<1 [a]	19 [a]	18 [d]	35 [d]	variable
Cucumber slices	5°C	1 [c]	12 [c]	70 [d]	14 [d]	marginal
Melon cubes	5°C	11 [c]	7 [c]	46 [d]	8 [d]	good/variable
Strawberries	5°C	17 [b]	5 [b]	51 [d]	19 [d]	good
Pineapple cubes	5°C	13 [c]	6 [c]	58 [d]	7 [d]	good/variable
Sliced mixed peppers	5°C	8 [b]	16 [b]	38 [d]	37 [d]	very good
Sliced leeks	5°C	11 [b]	15 [b]	<1 [d]	63 [d]	good
Swede cubes	5°C	<1 [a]	15 [a]	36 [d]	38 [d]	marginal
Frisee lettuce	5°C	<1 [a]	6 [a]	48 [d]	12 [d]	marginal
Spinach	5°C	18 [b]	3 [b]	36 [d]	19 [d]	marginal

a   Industry standard low O$_2$ MAP: 5% O$_2$/95% N$_2$ in OPP film
b   Industry standard: in air in microperforated OPP film
c   Industry standard: in air in OPP film
d   High O$_2$ MAP: 80% O$_2$/20% N$_2$ in PVDC-coated OPP film

© CCFRA 2001

High oxygen MAP for fresh prepared produce

Table 5: Summary of fresh prepared produce trials (Oct. 1997 – June 1999)

Prepared Produce Trial	Source / Variety	Harvest to pack time	Pre-pack treatments	Gas analyses (after 7 days at 8°C) Industry standard % O$_2$	% CO$_2$	High O$_2$ MAP % O$_2$	% CO$_2$	Overall achievable shelf-life (days) at 8°C Industry Standard	High O$_2$ MAP
Sliced mixed peppers	Dutch "California Wonder"	4 days	chlorine wash 6 mm cut	4 [b]	26 [b]	25 [c]	39 [c]	2 [b]	2 [c]
Iceberg lettuce #14	Spanish "Saladin"	4 days	chlorine wash 40-70 mm cut	2 [c]	14 [c]	51 [c]	17 [c]	4 [a]	11 [c]
Iceberg lettuce #15	Spanish "Enza Leopard"	3 days	chlorine wash 40-70 mm cut	1 [a]	10 [a]	48 [c]	15 [c]	2 [a]	7 [c]
Iceberg lettuce #16	Spanish "Salinas"	4 days	chlorine wash 10 mm cut	<1 [a]	11 [a]	40 [c]	23 [c]	2 [a]	4 [c]
Iceberg lettuce #17 (bulk trial)	Spanish "Salinas"	4 days	chlorine wash 10 mm cut	<1 [a]	27 [a]	<1 [c]	58 [c]	1 [a]	2 [c]
Cos lettuce #2	Spanish "Lobjoits"	7 days	chlorine wash whole leaves	1 [a]	9 [a]	50 [c]	15 [c]	3 [a]	7 [c]
Potatoes #5 (abrasion peeled)	British "Maris Piper"	7 months	sulphite dipped	2 [a]	10 [a]	25 [c]	31 [c]	1 [a]	4 [c]
			non-sulphite (EPL) dipped	1 [a]	15 [a]	9 [c]	46 [c]	1 [a]	1 [c]
Strawberries #2	British "Elsanta"	1 day	unwashed	21 [g]	0 [g]	32 [c]	19 [c]	4 [g]	4 [c]
Strawberries #3	British "Rhapsody"	3-7 hours	unwashed	21 [g]	0 [g]	35 [c]	21 [c]	1 [g]	4 [c]
				16 [b]	7 [b]	35 [c]	21 [c]	2 [b]	4 [c]

© CCFRA 2001

High oxygen MAP for fresh prepared produce

Table 5 (continued)

Prepared Produce Trial	Source / Variety	Harvest to pack time	Pre-pack treatments	Gas analyses (after 7 days at 8°C) Industry standard % O_2	% CO_2	High O_2 MAP % O_2	% CO_2	Overall achievable shelf-life (days) at 8°C Industry standard	High O_2 MAP
Curly Parsley	British "Bravour"	3-7 hours	chlorine wash trimmed	20[f]	1[f]	52[c]	14[c]	9[f]	7[c]
Little Gem lettuce	British	3 days	chlorine wash 35mm cut	17[b]	6[b]	52[c]	14[c]	7[b]	7[c]
Sliced mushrooms #2 (35-45mm white closed cap)	British "*Agaricus bisporus*"	1 day	chlorine wash 3 mm cut	<1[a]	9[a]	33[c]	19[c]	6[a]	8[c]
			unwashed 3 mm cut	12[b]	7[b]	12[c]	29[c]	2[b]	6[c]
				11[b]	9[b]	11[c]	37[c]	6[b]	>12[c]
Sliced bananas (ethylene ripened)	Hondurian "Cavendish"	19 days	Unwashed, peeled 6 mm cut non-sulphite (EPL) dipped	8[a]	9[a]	52[c]	13[c]	2[a]	4[c]
Sliced spring onions	Kenya	4 days	Trimmed, chlorine wash, sliced into 5-7mm	0[a]	36[a]	0[c]	29[c]	4[a]	5[c]
Baby leaf spinach	Spain	~5 days	Trimmed, Chlorine wash	18[b]	3[b]	55[c]	16[c]	7[b]	9[c]
Raddichio lettuce	Italy	~5 days	Chlorine wash, 35mm cut	2[a]	9[a]	63[c]	15[c]	3[a]	4[c]
Lollo Rossa lettuce	Spain	~6 days	Chlorine wash, 35mm cut	4[a]	9[a]	67[c]	12[c]	4[a]	7[c]

© CCFRA 2001

High oxygen MAP for fresh prepared produce

Table 5 (continued)

Prepared Produce Trial	Source / Variety	Harvest to pack time	Pre-pack treatments	Gas analyses (after 7 days at 8°C) Industry standard % O$_2$	% CO$_2$	High O$_2$ MAP % O$_2$	% CO$_2$	Overall achievable shelf-life (days) at 8°C Industry standard	High O$_2$ MAP
Red Oak Leaf lettuce	Spain	~6 days	Chlorine wash, 35mm cut	1[d]	11[d]	61[c]	13[c]	2[d]	2[c]
Flat Leaf Parsley	British	~3 days	Chlorine wash, 35mm cut	0[a]	12[a]	42[c]	16[c]	4[a]	9[c]
Cubed Swede	Scottish	2 days	Peeled, chlorine washed and cubed	7[b]	16[b]	27[c]	33[c]	3[b]	10[c]
Coriander	British	Same day	Unwashed	21[f]	0[f]	49[c]	13[c]	4[f]	7[c]
				0[a]	9[a]	49[c]	13[c]	4[a]	7[c]
Galia Melon	Moroccan	~9 days	Cubed	9[b]	22[b]	34[c]	30[c]	2[b]	2[c]
Raspberries	British "Glen Moy"	1 day	Unwashed	21[g]	0[g]	37[c]	29[c]	7[g]	9[c]
				17[b]	8[b]	37[c]	29[c]	5[b]	9[c]
Strawberries #4	British "Elsanta"	1 day	Unwashed	21[g]	0[g]	29[e]	29[e]	6[g]	6[e]
				21[g]	0[g]	29[h]	22[h]	6[g]	9[h]
				17[b]	6[b]	27[c]	40[c]	8[b]	6[c]

a Industry Standard low O$_2$ MAP: 5-8% O$_2$/92-95% N$_2$ in OPP film
b Industry Standard: in air in microperforated OPP film
c High O$_2$ MAP: 80% O$_2$/20% N$_2$ in OPP film
d Industry Standard: 5-8% O$_2$/92-95% N$_2$ in Polinex OPP film
e High O$_2$ MAP: 80% O$_2$/20% N$_2$ in Polinex OPP film
f Industry standard: in air in snap-lid tray
g Industry Standard: in air in open PVC punnet
h High O$_2$ MAP: 80% O$_2$/20% N$_2$ in Polinex OPP film + ATCO CO$_2$ absorber

© CCFRA 2001